GCSE
Resistant Materials
Technology
for OCR

Douglas Fielding-Smith

RECOGNISING ACHIEVEMENT

Heinemann Educational Publishers
Halley Court, Jordan Hill, Oxford OX2 8EJ
Part of Harcourt Education

Heinemann is the registered trademark of
Harcourt Education Limited

© Douglas Fielding-Smith, 2002

First published 2002

07 06 05 04
10 9 8 7 6 5 4 3 2

British Library Cataloguing in Publication Data is
available
from the British Library on request.

ISBN 0 435 41700 2

Designed and typeset by Artistix, Thame, Oxon
Index by Indexing Specialists (UK) Limited, East
Sussex. www.indexing.co.uk
Original illustrations © Harcourt Education Limited, 2002
Illustrated by Artistix
Printed in the UK by Bath Colour Books
Picture research by Peter Morris

Acknowledgements
The publishers would like to thank the following
students from St Paul's RC School, Leicester, for
allowing the use of their exemplar material (in
alphabetical order): Brendan Archdeacon, Louisa Atkin,
Simeon Bagshaw, Rhiannon Evans, Jennie Goulding,
Maciej Krychowski, Anne-Marie Liszczyk, Rachel Panter,
Sean Payne, Moira Quinn, Elizabeth Spencer, Myron
Sywanyk and Vicky Taylor.

The publishers would also like to thank the British
Standards Institution (BSI) for permission to reproduce
the BSI Kitemark and the CE marking on p.30, and
Barry Lambert for the table on p.77.

The author would like to thank: Georgina Royle;
Swanshurst School, Birmingham; and Sara Wood.

The publishers would like to thank the following for
permission to reproduce photographs: Advertising
Archive p.29; Corbis pp.78, 109, 120; Giles Chapman
Library pp.114, 115; Hemera Photo-Objects p.118;
Peter Morris pp.15, 24, 26, 28, 38, 39, 65, 79, 84, 86,
87; Techsoft p.56; TRIP/H Rogers p.112; John Walmsley
pp.67, 110.

Every effort has been made to contact copyright
holders of material reproduced in this book. Any
omissions will be rectified in subsequent printings if
notice is given to the publishers.

Tel: 01865 888058   www.heinemann.co.uk

Contents

Introduction

This book has been written to meet the specification requirements for the OCR GCSE in Design and Technology (D&T): Resistant Materials Technology. The book covers all the requirements for the short and full course; extra material and exemplification is contained within the Teacher's Resource File. The OCR specification is designed to meet the National Curriculum Orders for D&T and the GCSE Subject Criteria for D&T.

The programme of study for D&T at Key Stage 4 requires you to develop your D&T capability by applying knowledge and understanding when developing ideas, planning, making products and evaluating products.

The OCR specification content provides opportunities for you to develop D&T capability through activities, including:

● product analysis
● focused practical tasks that develop a range of techniques, skills, processes and knowledge
● design and make assignments, which include activities related to industrial practices and the application of systems and control.

You will be assessed in two ways. You will have internal assessment on your coursework (60% of the marks). You will also have to do a written examination at the end of the course (40% of the marks).

How to use the book

The book is divided into the following parts:

● Designing and Making
● Product analysis for the design specification
● Design thinking
● Product development
● Planning and production
● Product evaluation
● Making the grade
● Examination advice.

Within each of these parts, the knowledge and understanding of Resistant Materials Technology is covered, as well as help and advice on internal assessment objectives (coursework). ICT, industrial practice and health and safety are also covered.

The book is written in a series of double-page spreads which include:

● specification links – these show which sections of the specification are covered by the spread
● activities – these test knowledge and understanding and can be used in independent study and for revision
● key points – these provide a summary of some of the most important points on the spread and will be useful for revision.

Symbols are used on the spreads to show work covering ICT and industrial practice:

indicates ICT

indicates industrial practice.

At the end of each part, there is a set of more detailed questions that test your knowledge and understanding of the specification content.

The book is supported by a Teacher's Resource File, which provides more information on certain topics and proformas for coursework. Your teacher will let you have the sheets you need.

Resistant Materials Technology is an ever-changing subject and, as a result, many types of resource are needed to provide research and up-to-date information. The Internet is a valuable support in searching for help.

Notes for teachers

The OCR GCSE in Design and Technology: Resistant Materials Technology allows candidates to acquire and apply knowledge, skills and understanding through:

- analysing and evaluating products and processes
- engaging in focused tasks to develop and demonstrate techniques
- engaging in strategies for developing ideas, planning and producing products
- considering how past and present design and technology, relevant to a design and make context, affects society
- recognizing the moral, cultural and environmental issues inherent in design and technology.

Assessment objectives

Within this specification candidates will need to demonstrate their ability to:

- develop, plan and communicate ideas
- work with tools, equipment, materials and components to produce quality products
- evaluate processes and products
- understand materials and components
- understand systems and control.

The GCSE Subject Criteria (QCA 2000) sets out three specification assessment objectives for the scheme of assessment:

- AO1 Capability through acquiring and applying knowledge, skills and understanding of materials, components, processes, techniques and industrial practice.
- AO2 Capability through acquiring and applying knowledge, skills and understanding when designing and making products.

- AO3 Capability through acquiring and applying knowledge, skills and understanding when evaluating processes and products; and examining the wider effects of design and technology on society.

Examination

The terminal examination papers will test candidates' specialist knowledge, skills and understanding of Resistant Materials Technology through questions on the subject content outlined in the specification.

Internal assessment (coursework)

The specification assesses QCA's three assessment objectives in an integrated way through the following six internal assessment objectives:

- identify a need or opportunity that leads to a design brief
- conduct research into the design brief which results in a specification
- generate possible ideas for a solution
- develop the product for manufacture
- plan and realize the product
- evaluate and test the product.

DESIGNING
AND MAKING

Investigating a design problem

A design brief is a short clear statement of what you, as a designer and maker, intend to do. Before writing a design brief you will need to carry out an investigation. This will help in developing a clear and precise design brief for a marketable final product. You will first need to have a situation from which you can write the brief.

The situation

The situation is also known as the context or design task and will be the starting point from which you carry out your investigation. Your teacher may give you the situation or you may want to devise your own.

Here are examples of typical situations:

- Mechanical toys continue to provide interest, humour and fascination for all ages.

- Body adornments continue to be a form of self-expression and individuality.

- An unusual money storage system may encourage children to save their pocket money.

Analysis

An analysis of the situation will help you to understand what needs to be investigated; you will therefore have to ask yourself a number of questions which relate to the situation. This can be shown in a mind map analysis.

The types of questions you will ask yourself will obviously depend on the situation. Typical questions asked are:

- Who would the users be?

- Why would they need this product?

- Where would the product be used?

- What lifestyles do the users lead?

- What similar products are available on the market?

- What would the user want from this product?

- What are the selling points in each existing product?

- How often will the product be used?

- How much would the user spend on the product?

- Do the existing products cater for the user?

- What are the differences between each existing product?

- Are there experts or professionals from whom you could seek advice?

- Are there any popular trends or themes relating to the user or the existing products?

Once you have completed a mind map analysis of these questions you will then need to answer them. You will be able to answer some of the questions yourself using your own experience, but other questions will have to be investigated and researched. You will now have to ask yourself, where and how am I going to find the answers to these questions?

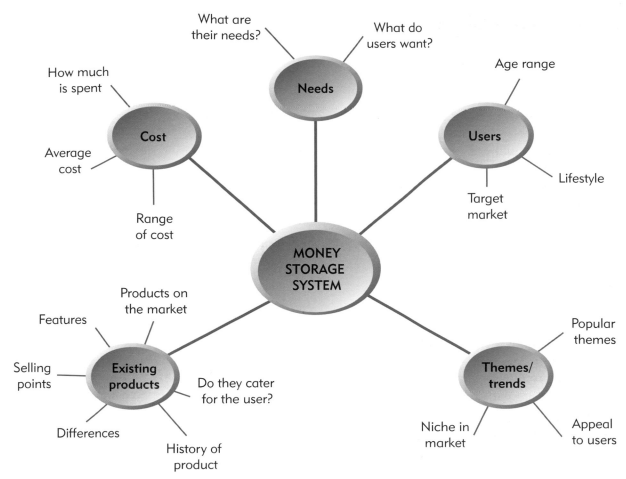

Mindmap analysis

Activities

1 Why do you need to carry out an investigation before you write a design brief?

2 How does a design brief help the designer?

3 Carry out a mind map analysis for:

- the chair you are sitting on
- the desk you are sitting at.

Key points

- A design brief is a clear and simple statement of what you intend to do.
- A situation is the environment in which the product will need to function.
- In carrying out an analysis you will be able to ask questions that will be investigated so you will be able to write a clear design brief.

The user

In order to design a successful product, you will need to find out people's opinions and what they as users need or want from a product. This can be done by writing a questionnaire and carrying out a user survey.

Questionnaires

Before writing a questionnaire you will need to consider the following questions:

- Who am I asking the questions to?
- What information do I want to get back?
- Will the questions identify a need?
- How will the questionnaire help me in writing a design brief?

When writing a questionnaire you need to make sure it is easy to understand, simple to answer and not too long. Present the information neatly by using word processing or software specifically designed for surveys. Always start with an introductory paragraph such as:

My name is _____ and I attend _____ School. For my resistant materials coursework I am _____

At the end of your questionnaire thank the participant, for example:

Thank you for taking the time to complete this questionnaire

Two types of questions can be undertaken, qualitative and quantitative.

- Qualitative questions ask for an opinion. The response to these types of questions needs to be written down as the participant answers you, for example:

What do you think of the mechanical toys available to buy in the high street shop?

- Quantitative questions require a response that can be measured. For example a box can be ticked or a grade can be given. These answers can be shown in a graph or chart, for example:

Do you think a mechanical toy should be educational?

Yes ☐   No ☐   Don't know ☐

Rate the following features of a mechanical toy on a scale from 1 to 5 where 1 is not important and 5 is very important.

Ease of use ☐   Material ☐   Cost ☐

Tips

- When asking questions with a tick list always give an 'other' or 'don't know' option just in case you have not included everything.
- You should aim to have at least 10 questions with only two qualitative questions included.
- To get a good cross section of opinion try to get the questionnaire completed by at least 15 people.

Once you have all the completed questionnaires you will need to display the results for each answer in the form of suitable bar charts, pie charts or graphs. This can be done using software such as a spreadsheet package.

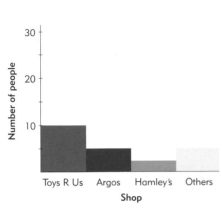

A bar chart showing the result of a questionnaire

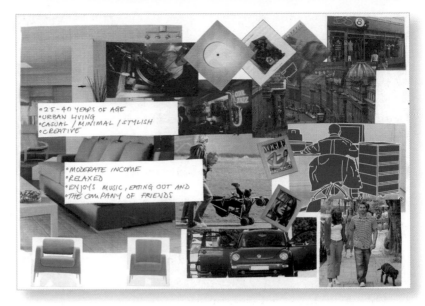

A user profile

User lifestyles and activities

Information on how the user lives and what activities they participate in will give a good idea of what they need. This will help you to identify needs and opportunities for a design brief. When carrying out a user questionnaire you may want to include questions that relate to their lifestyle.

Examine magazines and publications that would be read by the user and look out for:

- trends (areas of interest)

- activities (shops they use)

- styles (types of products that appeal to them).

From this information, your own observations and suggestions from your friends and family, a user profile can be created. This will be presented in a combination of pictorial and written form, for example pictures from magazines or the Internet and a summary of the major points. This will help you understand what type of people will use the product.

In your user profile try to show the following:

- The likes and preferences of the user.

- What they do for a living and what they do in their spare time.

- The type of house they live in or what kind of job they have.

- What they might look like, the clothes they wear and the cars they drive.

Activities

1 What is the difference between a qualitative and a quantitative question?

2 Look through magazines and collect pictures that could be used to profile a female aged 15. Present your findings on A3 paper and add text to build up a complete user profile.

Key points

- A questionnaire is used to gain first hand opinions from the user.

- A user profile is used to give you a better idea of the type of person who would use the product.

The market

Before you start designing you should examine the market to see if there are any similar products available for the user to purchase. This will provide you with information about the following.

- The range of existing products available.
- The type of user the product is aimed at.
- The price difference.
- The difference between each product in style and function.

You can carry out such an investigation by:

- visiting shops
- collecting retail catalogues
- visiting websites.

Collecting and recording

Pictures and photographs, along with information about the product, can be laid out on a page. This will give a cross section of the products available. You can then analyse the difference between these products in a written report. In doing so you may see that there are potential market possibilities. In the conclusion of your report try to identify where your design may fit in. Ask yourself the following questions:

- Are any of the products not aimed at certain user groups?
- Can a more inexpensive product be produced?
- Is there scope for a better functioning design?
- Is there scope for a more stylized design?
- Are there any design trends or themes?

- Black lacquered steel with glass top
- Storage shelf
- W118 x D78 x H48 cm
- £85

- Lacquered beech or birch veneer
- W90 x D55 x H45 cm
- £20 each

- Lacquered solid pine
- W55 x D55 x H45 cm
- £15

- Plastic melamine
- Storage shelf
- Mobile (castors)

- Glass top
- W50 x D50 x H56 cm
- £55

Existing products available

Case study   *The clockwork radio*

While watching television, Trevor Bayliss saw a programme about the spread of AIDS in Africa. The programme highlighted the fact that it was difficult to broadcast educational information about the disease as many parts of Africa have no electricity and many people cannot afford to buy batteries for their radios.

Trevor remembered that at the beginning of the last century, British colonialists in Africa used to listen to records on wind-up gramophones. Trevor thought that if record players used to be driven by a spring attached to a dynamo, then why not have a radio powered by the same method?

He set about making a radio that gives 40 minutes of play from 20 seconds of winding. He had difficulty finding a manufacturer to produce the radio, but after Trevor appeared on *Tomorrow's World*, a South African company began making the radio. In June 1996 the Freeplay radio was awarded the BBC design award for best product and best design.

A clockwork radio

The demand for the self-sufficient 'wind up' technology has increased. Trevor has applied the same idea to a number of products such as electric generators, lights, mobile phones and laptop computers. He is now working on the idea of shoes that build up electrical power as you walk!

Advice from experts

Advice from experts can assist in identifying problems and user needs. Experts are people who have experience in and know a lot about the area you are investigating. They may offer help in the research and development of your product that you cannot find anywhere else. Some of these experts might include: designers; manufacturers and producers of a related product; engineers; psychologists; researchers; ergonomists; and scientists.

You may not be able to speak to these people directly, but you can source information written by them in books, magazine articles and on the Internet. If you do get the opportunity to interview someone, this can be done by face-to-face questioning, by sending a letter/questionnaire or by e-mail. You can present the information you receive in a written report or graphic display.

Activities

1 What are potential market possibilities?
2 How could talking to a designer help in your investigation?
3 What design need did Trevor Bayliss see?
4 Name two other products that would benefit from wind-up technology.

Key points

- In examining existing products on the market you may be able to see potential market possibilities.
- Experts can offer important information and guidance in writing a design brief.
- People's needs can be addressed by using and adapting existing technology.

Writing the design brief

At the beginning of your investigation you will have brainstormed the questions and areas that needed to be looked at in order to design your product. Answers to these questions will have been shown in the different areas of investigation you have carried out (such as those outlined below). You will need to analyse the information you have collected to form a conclusion. This will help identify a need and help you write the design brief.

Analysis of the investigation

Questionnaire

You will have obtained basic raw data from conducting a user survey and then displayed the results in graphs. A computer database is a helpful tool in analysing this data. You will need to set up a record for each person who answered the questionnaire with each individual question called a field.

Once the data has been stored you will be able to sort it into a required order and search for information. For example, 'How many fifteen year old girls who own a mobile phone own a Nokia?' or 'How many people with children under five shop at Toys R Us?' This will give you a better idea of:

● what the user looks for in a product

● what the user wants from a product

● their opinions.

User profile

In producing a profile of the user you will have a better idea of:

● who the user is

● what appeals to them

● what lifestyle they lead.

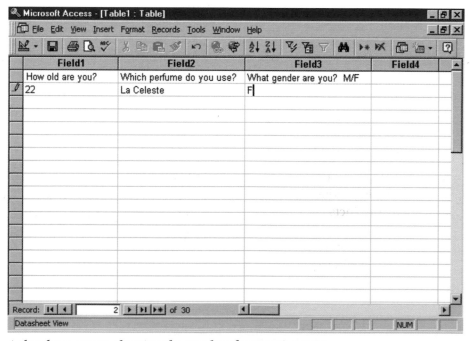

A database screen showing the results of a questionnaire

Existing products

In examining the range of products on the market you will have a better idea of:

- the style, function and price of similar products
- the users they are aimed at
- the market possibilities.

Advice of experts

In seeking the advice of experts you will have a better idea of:

- how they view problems
- technical matters.

Identifying the need

You will need to write down what you have found out. This written conclusion will highlight the needs and design opportunities you have found during your investigation. Use evidence from your investigation to justify your conclusions on:

- the user
- the user's needs
- opportunities for designs.

The following are examples of what you may find out through your investigation:

- Looking at the existing products available, you may have discovered there are no products aimed at teenagers.

- When questioning users you may have found out that some products are too expensive for a large percentage of the people interviewed.

- When questioning manufacturers they may have expressed a difficulty in producing environmentally friendly products.

Think about how these needs can be combined with the needs of the following:

- the community
- your interests
- the home
- industry.

An example design brief

When writing the design brief try to keep it short and precise. Start with, 'I am going to design and make _____' You do not need to be too detailed because you have not researched or started to design the product yet. If you add too much detail this will restrict what you can do.

The situation

Mechanical toys continue to provide interest, humour and fascination for all ages.

The conclusion of the investigation

I have found out that mechanical toys are either expensive collectors' pieces or are inexpensive toys for young children. Therefore most mechanical toys are designed with these users in mind. I have found out through my survey that teenagers would buy mechanical toys if they were affordable and appealed to them.

The design brief

I am going to design and make a mechanical toy that appeals to teenagers and is affordable for them to purchase.

Activity

1 Write a design brief for:
- your desk
- your chair.

Key point

- To write a design brief you need to analyse all of the investigation work you have carried out.

17

Questions

As a designer you have been given the following design task (situation) to investigate:
An unusual money storage system that may encourage children to save their pocket money.

1 Draw a mind map to analyse the situation.

2 To find more information about the user a survey must be carried out.

   a Write five quantitative questions that could be asked.

   b Write two qualitative questions that could be asked.

   c Explain the difference between a qualitative and quantitive question.

   d How will carrying out a user survey help with writing a design brief?

   e State three ways in which the results of the questionnaire can be shown.

   f Name the software that could be used to analyse the results of the questionnaire.

3 In investigating the situation you will have to produce a user profile. Explain what a user profile is and why it is useful.

4 Investigating existing products can help identify market possibilities.

   a What type of shop or website would you visit to gain more information about the situation above?

   b How would you display the information you have collected?

5 Advice from experts can assist in identifying problems and needs.

   a Name two experts that would be useful to talk to about the situation.

   b If you could not talk to them how would you source the information?

PRODUCT ANALYSIS FOR THE DESIGN SPECIFICATION

Product analysis past and present

In Section 1 you have investigated a situation which resulted in the writing of a design brief. In this section you are going to research the design brief. This process involves finding out information and analysing it so that conclusions can be made. This will help you write a specification for the design and evaluation of your product.

Product analysis

Examining products from the past and the present day will allow you to see how and why products have changed and evolved. The products you study should relate to your brief. To find information on them you will need to look in books, magazines and on the Internet. The information and pictures you find can be recorded in a time line. This will show how and why the product has changed chronologically. For each stage write down how and why the product has changed. Make sure you include the following information:

The aesthetics

- How has the overall shape changed?
- Have trends or styles influenced its shape?

The users

- What did the user at that time want from the product?
- How have user needs changed?

Materials and processes

- What materials were used to make the product?
- What manufacturing processes were used?

- What effects did new materials and processes have on the cost, shape, quality and function of the product?

You should also collect pictures and information about present day products that relate to your brief. Catalogues are the easiest way of researching this, for example:

- inexpensive to expensive
- different age ranges
- different materials.

If you are aiming at a specific area such as teenagers then you will need to collect pictures that relate to this market.

Once you have collected the pictures and information you can then evaluate the products. This can be done visually by using star diagrams (see page 22), tables and written evaluations.

Activities

1 How have users' needs changed over the last 50 years? Try to give reasons for your answers.

2 Apart from the shape, name two other ways products have changed over the last 50 years.

Key point

- Products have changed and evolved over the years because:
  - the users' requirements and needs have changed
  - technology has advanced.

1925–27

**Marcel Breuer
Wassily chair**

- Bent tubular steel support and fabric sections
- Revolutionary use of tubular steel and method of manufacture for the time
- Utilised industrial techniques to produce the chair for large volume production

1951–52

**Arne Jacobson
Ant chair**

- Plywood seat and bent tubular steel legs
- Easy construction with two elements – seat and legs
- Design allows for mass production
- Inexpensive to produce
- Design later developed to allow the chair to be stacked and easily stored

1962–63

**Robin Day
Polyprop chair**

- Injection-moulded polypropylene seat with bent tubular steel legs
- Low cost, stackable seat using a new material for the time
- Polypropylene is inexpensive, durable and lightweight
- Injection-moulded seat can be quickly and inexpensively produced

1979

**Giandomenico Belotti
Spaghetti chair**

- Tubular steel frame and PVC strips forming seat and back
- Strong structure and can be stacked
- PVC strips shape to seated user, for comfort
- A clever and inexpensive solution for the seat and back

1988

**Philippe Starck
Dr. Glob chair**

- Tubular steel frame with polypropylene seat
- A strong, stackable chair
- Produced using mass production methods, but unique and elegant design

Analysing products from the past

Recording your analysis

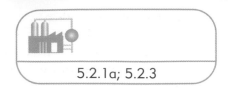

When recording your product analysis you will need to decide on certain criteria against which to analyse the product. The criteria outlined below can be used as a guide in analysing products.

Criteria checklist

User and target market

- What is the market for the product and how do you know this?
- Who uses it/buys it?
- Does the product fulfil the user's needs and requirements?

Function

- Has the designer taken the user into account?
- Does the product look easy to use?
- What ergonomic considerations have been made?

Material

- What materials/finishes have been used and why?
- What properties does the material used have?
- Does the method of manufacture affect the chosen material?

Cost

- Does the cost of the product affect the chosen material?
- Is the cost affected by the market the product is aimed at?
- In comparison with similar products, is the product expensive?

Aesthetics

- Has the designer used a theme or style, and if so, why?
- How has the designer used colour/material/finishes?
- How has the designer used proportion/balance/symmetry?

Star diagrams

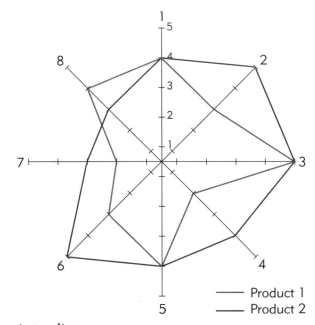

A star diagram

Using a star diagram is a simple and effective way of evaluating both your own ideas and existing products. You will first need to decide on the criteria you will judge the products by. You should be able to grade these criteria from 1 to 5, for example, aesthetics, cost etc.

On each point of the star there are numbers which correspond to the criteria. Each line of the star is numbered 1 to 5 where 5 is excellent and 1 is poor. You need to evaluate the product and award it a mark from 1 to 5. Once you have done this for every criteria point you can then join up the marks to create a star shape.

If you use the same criteria to evaluate several products you can easily see which is better. This can be done using a star diagram for each product or by evaluating all the products on the same star using a different colour for each. Make sure you remember to add a colour key to help the viewer interpret the information.

Written evaluations

Your star diagrams should be used in conjunction with a written evaluation. This will enable you to justify the mark you have given to each product. You can also add extra comments in the written evaluation using your criteria as a guide. Carrying out this kind of research will allow you to examine the intended purpose of the product. In your conclusion try to sum up the following:

- good points
- bad points
- improvements you could make to the product.

Activity

1 Evaluate a product using a star diagram. Ask a friend to evaluate the same product using the same criteria. Compare your results. How do your opinions differ?

Key points

- Designers have to consider a wide range of issues when designing a product.
- In analysing existing products you will be able to see how designers have addressed these issues with both good and bad solutions.

Using existing products

When you begin designing you are not expected to re-invent the wheel. By examining a range of existing products, similar to that which you wish to design, you will have first hand knowledge of how the product functions and how it is constructed.

Examining existing products

You will first need to collect a range of similar products. If this is not possible you may have to visit a shop that will have a range of similar products. Try to use the product as it was intended and evaluate it as both a user and designer.

The following evaluation sheet can be used as a way of reporting what you find out.

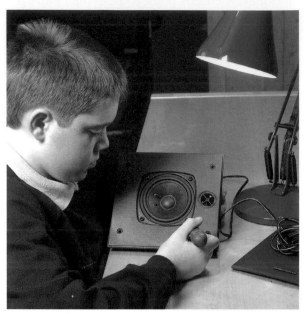

A student using a product to evaluate it

Product evaluation

Product: _____

Name: _____

Manufacturer: _____

Cost: _____

Materials used: _____

Fitness for purpose: _____

Finish: _____

Product functions: _____

Extra features: _____

Intended market: _____

Quality and issues: _____

Aesthetics: _____

Maintenance: _____

Ease of use: _____

Summary: _____

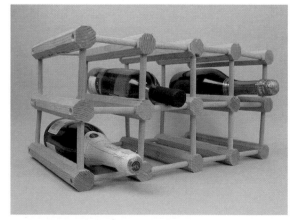

A sketch or photo of the product will help explain the evaluation

The terms on the evaluation sheet are explained below:

- Product: What the product is.

- Name: The name of the product or model (if it has one).

- Manufacturer: The name of the company who made the product.

- Cost: How much the retail cost of the product is.

- Materials used: Name the materials used in the product which you can identify. Try to be as precise as you can, e.g. beech, stainless steel, polypropylene.

- Fitness for purpose: Are the materials used appropriate for their purpose? Would any other materials be better?

- Finish: Has the product been painted, varnished or coated? Is this finish appropriate? Could a more appropriate finish have been used?

- Function: What does the product do? Try to be specific and say exactly what it is used for. Does it function well?

- Extra features: Does the product have any special functions or features that other products do not have?

- Intended market: Who is the product aimed at? What type of person would use it?

- Quality: Has the product been well made? Has it been finished well? Has the product been assembled well?

- Issues: Is it safe to use? What has or has not been done to make it safe? Can it be recycled or does it use recycled materials? Has it used more material than is needed? Could it use less material or different materials to make it more environmentally friendly?

- Aesthetics: Does the product have a stylish look? Does it follow any trend? Has it been over-styled or under-styled? Does it contain any graphics?

- Maintenance: Could the product be repaired easily? Can it be cleaned easily?

- Ease of use: How easy is the product to use, hold, turn on, work etc.?

- Summary: Sum up all you have found out to give a final conclusion. Add any good and bad points that are worth mentioning or any extra comments if necessary.

Once you have completed an evaluation of a range of similar products you will be able to compare them with one another. This will enable you to see good and bad examples of products. This will influence your thinking with regards to designing and making your chosen product.

Activity

1 Use the information in this section to carry out a product evaluation of your own.

Key points

- In examining a product first hand you will be able to evaluate it in greater detail.

- In evaluating the product in use you will be able to see how it functions and relates to the user.

Disassembling existing products

To get a better understanding of how a product has been designed and manufactured you will need to disassemble an existing product. This is done in three stages:

- Taking the product apart.
- Recording, through photographs and sketches, how the parts go together and relate to one another.
- Evaluating the disassembled product.

Taking the product apart

In taking the product apart you must be careful to note where and how the parts fit together. You will have to record how the parts relate to one another. More importantly you will have to re-assemble it after you have taken it apart. The following points should be borne in mind when disassembling a product:

- Ensure the product has no power running to it.
- Locate the fixings (screws, nuts, bolts, clips, etc.) and decide which tools are needed to remove them. The fixings may sometimes be hidden by a removable cap or cover.
- Collect the correct tools for removing the fixings and clear a space for the activity to take place. A variety of pots and bags will be useful to place small parts and items in so they do not get lost. Use masking tape or self adhesive stickers to label components for easy re-assembly.
- When removing fixings be careful not to force them. Some parts may not come apart because they have been glued or permanently fixed. Be careful not to dismantle these sections.
- Be extremely careful. Products can become dangerous when casings are removed and systems revealed.

Recording the disassembly

The easiest way to record the disassembly is by photographing each stage. This can be done by fully disassembling the product first and then putting it back together without fixings. The product can then be photographed as a whole. Remove the first piece, place it next to the product and photograph it again. This can be repeated until all the parts are laid out.

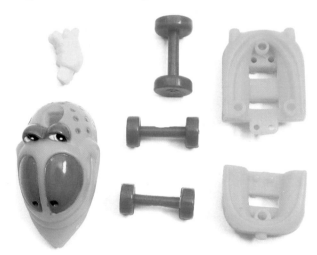

A disassembled product

Another method of recording the disassembly is through an exploded drawing. The advantage of this method is that parts which are glued or permanently fixed can be shown to be separated. It also allows you to show where all the separate parts fit in relation to one another and helps explain how the product is assembled.

Exploded view of a child's swing

Evaluating the disassembled product

Showing the disassembled parts in photographs or an exploded drawing enables you to label them and identify their functions. Once each part has been identified you can evaluate the product. The following chart can be used as a guide for doing this.

Product: _____

Number of parts: _____

Part number	Name	Material	Function	Process used

How the product works (including scientific principles): _____

Alternative methods and solutions:
_____

Testing: _____

The terms on the chart are explained below:

- Product: What the product is.
- Number of parts: How many parts are there in total.
- Part number: The number of the part identified in either the photographs or exploded drawings.

- Name: The name of the part, e.g. the outer casing, handle, top, bottom, etc.
- Function: The function of that part within the product.
- Material: The material the part is made of (try to be specific, e.g. beech wood, stainless steel, acrylic, etc.).
- Process used: The process or processes used to make that part (try to be specific, e.g. cold pressed and welded, vacuum formed, etc.).
- How the product works (including scientific principles): An explanation of how the parts relate to each other to make the product function. Diagrams can be used to help explain this. You may want to include scientific principles such as systems, mechanisms, electronics, pneumatics or structures to help explain how the product works.
- Alternative methods and solutions: Try to explain the different ways that could be used to make the product, e.g. different materials, parts, assembly, scientific principles or process of production.
- Testing: Comparing the materials and processes of production against other materials and processes. Try to offer alternative materials and processes that could be used.

Activity

1 Use the information in this section to disassemble a product and carry out an evaluation of your own.

Key point

- Disassembly of existing products will give you a greater understanding of how:
  - the product has been designed and manufactured
  - each part relates to one another
  - the product works
  - there are alternative methods that can be used.

The user

In analysing and evaluating existing products the needs and opinions of the user are important. Ask yourself the question 'What does the user think of the products available?' This will allow you to gain a wider view of how the products fulfil the user requirements.

Market surveys

You will need to carry out a market survey using a questionnaire (see page 12) to gain a better understanding of the user. The types of questions you will ask must relate to existing products on the market. This can be done by allowing the person being interviewed to give their opinions by:

● physically using the product

● examining pictures and information about existing products

● answering questions related to similar existing products they have used.

The questions you include in your questionnaire will need to relate to how the product is used. Considerations should also be given to any moral, social, economic, cultural and environmental issues that relate to the product.

Social and moral issues

These relate to how the user views the product ethically; how it relates to their views and opinions and to society as a whole. The questions you might ask could include:

● If the product was made outside the UK would you still buy it?

● Does it matter where the product is manufactured?

● If the manufacturer producing the product had bad environmental practices would you still buy the product?

Is the user concerned with where the product is manufactured?

- If the manufacturer producing the product treated the workers badly would you still buy the product?

- If another company producing the same product had a better work policy would you rather buy the product from them?

Economic issues

These relate to the cost of the product and its value for money. The questions you might ask could include:

- Is the product worth the retail price?

- How much would you spend on this type of product?

- Would you spend more if the product were built to a higher standard?

Environmental issues

These relate to how environmentally friendly the product and manufacturing processes are. The questions you might ask could include:

- Do you think the product is over-packaged?

- Does the product need any packaging?

- Would you prefer the product to be made from recycled material?

- Would you pay more for the product if it were made from recycled material?

- If the product were made from a non-recyclable material would this affect you buying it?

Aesthetic issues

These relate to the styles and how the product looks. The questions you might ask could include:

- Does the shape appeal to you?

- What range of colours would you like the product to be in?

- Are the graphics an influence when buying a product?

Will having a range of colours influence the user?

Activities

1 Why would questions about the use of a product be helpful in your research?

2 Why are questions about social and moral issues important?

3 a Take an existing product and examine it. Try to evaluate it from the following angles:
- an environmental view
- an economic view
- a social and moral view.

b After carrying out this evaluation would you still buy this product? Give reasons why you would or would not buy it.

Key points

- There are a wide variety of issues to consider before purchasing a product.

- Designers and manufacturers may not always consider the wider issues when producing goods.

Technical data

When designers and manufacturers produce a product they have to consider health and safety requirements. This ensures that the product will not harm the user, and in turn the user will know that the product they are buying will be of good quality, fit for its purpose and safe to use.

British and European standards

BSI (British Standards Institution) along with other international standards organizations has produced requirements for designers and manufacturers in a wide range of products. The designer should therefore use these standards as a guide when designing a product.

It is a legal requirement that products conform to a health and safety standard. Products can be tested by BSI and other organizations around the world. If a product meets the specific requirements of a Standard then the manufacturer may be awarded the BSI Kitemark, which may be displayed on the product.

In order that some products may be placed on the European market, many products must meet the requirements of specific European Directives. Having successfully met these requirements the product can carry the CE marking.

The BSI Kitemark

The CE marking

Product sizes

When designing products, along with considering human sizes, sizes of standard products are also important. For example, if you are designing a CD holder you will have to find the length, width and depth of a CD case.

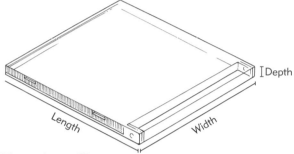

Measuring a CD case

There are a few standard sizes of products such as paper, batteries, etc., but you may have to find out the size yourself for your specific product. This will involve finding and measuring the product and designing your product around it.

The overall size of the product is therefore determined by these sizes. As part of your research you will have to collect and present this data. For the product to function correctly, it is crucial that these dimensions are sourced at the very beginning.

Once the basic data has been collected you will be able to use it as a starting point for your ideas. For example, if you were designing a CD holder you would have to think about:

- How many CDs are to be stored?
- How much space should you leave in order to remove and replace the CDs easily?
- How are the CDs to be stored – stacked up, across, etc?

Your initial ideas will reflect this research and give you a good starting point to begin designing. This relevant data should also be included in your specification.

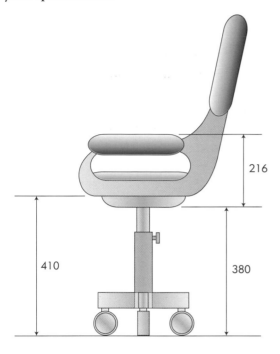

Standard sizes for a chair

Activities

1 Why are products tested?
2 Why are products given marks after they have been tested?
3 What size would a product have to be to hold:
  - 10 CDs?
  - 10 sheets of A4 paper?

Key points

● When producing a product that has to accommodate an existing product, it is important to research product sizes before creating the design.

● Standards are used to ensure that the product is fit to be used.

Ergonomic and anthropometric data

5.1.2b; 5.1.3e

An essential part of good design is that the product can be used comfortably and easily.

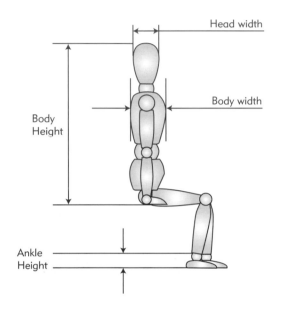

Ergonomics

Ergonomics is the study of a person in their environment and how they use or interact with products. When designing a product, considerations have to be made concerning the size and shape of the user. This so that the product is ergonomically correct.

Anthropometric data

Designers use anthropometric data to assist in designing their products. The data is collected from a large number of people and shows the average size of parts of the body. This will allow the designer to get an average size of a person. It also allows for sizes of specific groups in the population such as boys, girls, men, women and people of different ages. Therefore the product can be specifically designed for particular user groups.

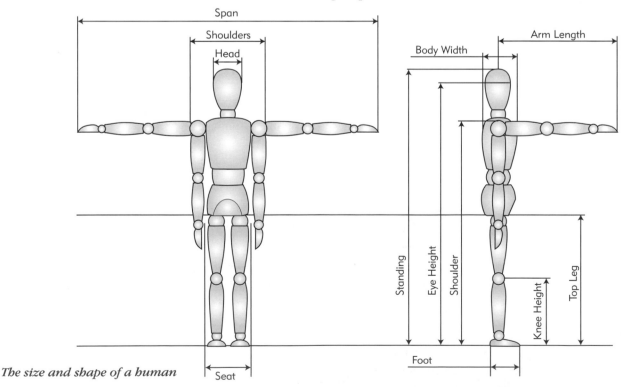

The size and shape of a human

The anthropometric data can also divide the sizes into 5 per cent, 50 per cent and 95 per cent of the population. For example, if you were designing a desk, the height of it would have to be correct for the majority of the population (95 per cent). This would therefore be the best size to base the design around, as it would suit most people.

Anthropometric data (all dimensions in mm)

		4 Years	8 Years	12 Years	14 Years	Adult
Shoulder Width	Female	246	282	330	363	399
	Male	239	290	330	386	455
Head (Ear to ear)	Female	137	142	145	147	147
	Male	142	145	147	152	155
Seat (Hips)	Female	196	231	285	325	362
	Male	188	229	269	323	335
Standing	Female	1039	1270	1478	1613	1605
	Male	1039	1270	1499	1709	1755
Eye height	Female	937	1168	1376	1511	1506
	Male	937	1178	1397	1607	1643
Shoulder height	Female	772	1008	1240	1326	1265
	Male	775	1011	1204	1410	1387
Knee height	Female	272	351	422	445	452
	Male	269	351	417	478	503
		4 Years	8 Years	12 Years	16 Years	Adult
Leg	Female	437	577	696	734	732
	Male	437	577	693	800	833
Arm length	Female	417	539	653	706	666
	Male	424	546	643	754	729
Foot	Female	165	196	216	239	244
	Male	168	196	218	249	267
Ankle height	Female	56	69	79	76	76
	Male	56	69	79	84	89
Body width (chest to back)	Female	132	137	160	175	–
	Male	132	145	163	188	229
Body height (head to seat)	Female	587	676	770	848	–
	Male	597	686	760	876	909

Activities

1 Look at the ergonomic table above. What is the best size for:

  ● the height of a chair (floor to seat)

  ● the width of a chair seat.

2 Compare these sizes to the size of a number of chairs in your classroom. Are all the sizes the same? Write down your conclusions.

Key point

● Correct ergonomics are essential for the design of products. Without the application of anthropometric data to your design the user will find the product awkward and uncomfortable to use.

Writing a specification

Having completed your research and analysed all of the information you will now have to conclude it by writing a specification. A specification is very important stage in the design process. You will refer to the specification throughout the project so you will need to get it right. It is important that you write a clear and detailed specification, which identifies all the criteria for the product to be successful.

What is a specification?

A specification is a list of what your product must do, have, be, include, etc. This will act as a guide when you are designing, planning and making. It will ensure that your product is fit for its purpose.

The specification will be criteria you will use when evaluating your work throughout, and at the end of, the project. You must be able to show, with supporting evidence, that you have met each specification point. For example, if you wrote, 'The product must look good' or 'The product must work well' how will you prove this?

Specification checklist

The following is a checklist that will help you in writing the specification. This should be used as a guide only; some of the headings may not apply to your project. Answer the questions in relation to your own project and research.

Remember:

- The research you have carried out should be used as your guide in deciding what to write.

- For each point, write sentences giving a reason why it is included. You may use your research to support these reasons.

- The product, or parts of the product, must be manufactured using a control device such as a jig, template, CAD/CAM system, etc. This is to ensure that the product can be manufactured in quantity. It will also assist in accuracy, which will enhance the product's consistency and quality.

- The specification should not predict what the product would look like. This will limit what you can do as well as making design decisions before any designing has taken place. Do not, for example, specify that your product 'must be painted red' or 'must be made from beech and mild steel'.

Time scale

- How many hours/lessons/weeks have you got to complete the entire project?

- What date is it due to be handed in?

The user

- What is your target market?

- What will the user want from the product?

Monetary cost

- Do you have a budget? If so, what is it?

- Will the final product be affordable for users in your target market to purchase?

Function

- What is the purpose of the design?

- What is it meant to do?

Reliability

- How will you ensure that the product will work every time?
- What is the product's life expectancy – how long will it last?

Aesthetics

- What aesthetic considerations should be taken into account to ensure that the product appeals to the user?

Performance

- How and where is the design meant to work?

Weight and size

- Are there any specific sizes that have to be considered?
- Is any consideration about weight required?
- What ergonomic and anthropometric data has to be used in the design?

Materials

- What qualities and properties will the materials have?
- Are there any limitations in the availability of materials?

Health and safety

- What health and safety factors are involved in the design and construction of the product?

Manufacture

- Are there any limitations in the availability of tools and machinery?
- Are there any limitations in your own ability and experience?
- How can the product be manufactured to ensure a quality outcome?

Batch production

- Why is a control device (e.g., jig, pattern, CAD/CAM, etc.) necessary in the manufacture of the product?

Moral, social, cultural and environmental issues

- What environmental considerations must be taken into account in the design, materials and manufacture?
- What moral, social and cultural considerations are there in the design and manufacture of the product?

Activity

1   Using the checklist, write a specification for:

   a   the pen you are using;

   b   the chair you are sitting on;

   c   the desk you are working at.

Key points

A specification is:

- a list of key features that your product must have;
- a guide for your design thinking;
- criteria by which you will judge your designs and final product.

Questions

1 Examining products from the past and the present is an essential part of research.

   a Why is this type of research useful to the designer?

   b Explain what is meant by disassembly of an existing product and why is this important.

2 Ergonomic and anthropometric data are often used in the design of products.

   a What is ergonomic data and how is it used in the design of products?

   b Explain the difference between ergonomic and anthropometric data.

3 When a Kitemark is placed on a product, what is it showing?

4 Describe what each of the following terms mean in relation to a product:

   a social and moral issues

   b economic issues

   c environmental issues

   d aesthetic issues.

5 For each of the following products, write down three specification points and give reasons for each. An example has been given for the first.

   a

   Specification point: The product must be durable.

   Reason: It has to withstand a variety of weather conditions.

   b

   c

   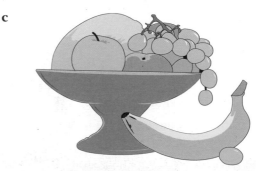

   d Explain the purpose of a specification.

DESIGN THINKING

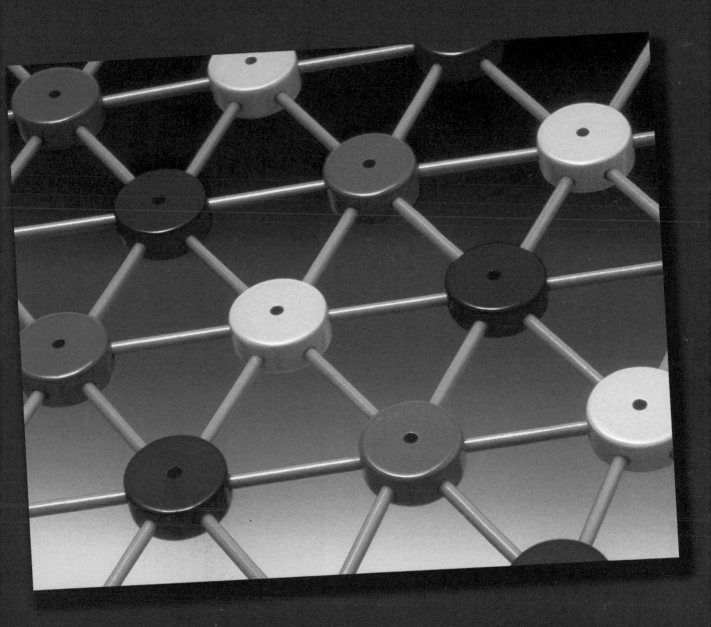

Communication of ideas – drawing

When you begin designing and developing your product it is important to communicate your ideas in a clear way. You may understand what you have done but others may not. Using a variety of methods to communicate your ideas will allow others to see your progression of thought. It will also allow you to clearly see what you have done, how it can be developed and how you can solve particular problems.

Drawn communication

There are two types of drawn communication:

- Informal: often used in the initial ideas and development stages as it allows ideas to be communicated quickly and easily.

- Formal (accurate drawings using specialist equipment): This is often used at the end of development and in the planning stages. This type of communication will allow you to accurately show your final ideas and produce working drawings from which the product can be manufactured.

Drawing equipment

To produce quick, informal sketches specialist drawing equipment is not used. For sketching use a well sharpened HB or 2B pencil so your sketch can easily be altered with an eraser. Colour and ink can be added after the sketch is correct. In order to transfer ideas quickly, layout or tracing paper can be used.

To produce accurate, formal drawings quickly, specialist equipment is used. For this type of drawing use a harder pencil that is well sharpened, such as H or 2H. When you begin drawing, make sure that the lines are drawn faintly. After the drawing is finished and correct you can then darken in the lines.

Drawing boards and tee-squares/parallel motion

The paper is mounted onto an A3 board using a clip or low tack tape. Ensure that the horizontal edges of the paper are parallel to the tee-square or parallel motion. When horizontal lines are drawn along the edge of the tee-square they will be parallel to each other and to the horizontal edges of the paper.

A drawing board and parallel motion

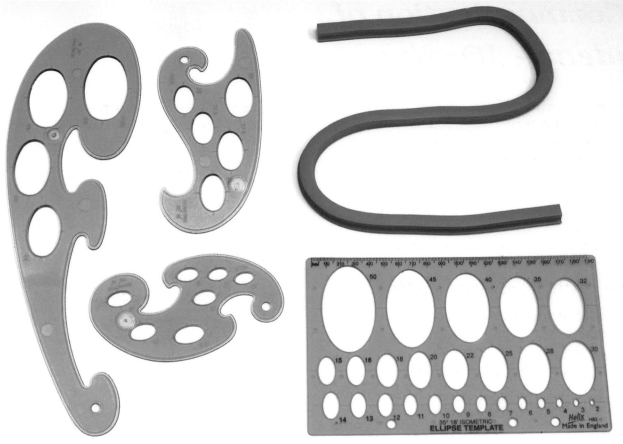

A French curve, flexi-curve and an ellipse template

Drafting aids

Templates are available for drawing circles and isometric ellipses in a range of sizes.

Other necessary equipment such as a compass, 300 mm rule and a set square will often be needed to complete a drawing. A set square can be placed on a tee-square to draw vertical lines that are 90° to the horizontal. The angles on set squares are either:

- 30°/60°/90° so it can be used to construct isometric drawings
- 45°/45°/90° so it can be used to construct orthographic drawings.

Where the curves on your drawing are irregular, French curves and bendable flexi-curves can be used.

Activities

1 Why are good communication skills required?

2 Why are there informal and formal drawings?

3 What is a:
- tee square/parallel motion?
- set square?

Key points

- There are two basic ways in which you can communicate your ideas, informal and formal.
- Drawing equipment is used with formal drawing to ensure accuracy.

Communication of ideas – 3D views

There are a wide range of methods through which you can communicate your ideas. By using these methods three-dimensional (3D) ideas can be communicated on paper. These methods can be used in both informal and formal ways.

Pictorial 3D views

The best way to show your ideas is by drawing them in 3D so that they look like how you would see them. This allows you to show a lot more information because you are seeing the top (plan), front and end surfaces of your drawing. It also allows you to apply tonal shading which creates a more realistic idea of what the product will look like.

Perspective views

Objects can be drawn in perspective, as you would see them in real life. This means that as the sides recede from the viewer they appear to get smaller. It allows objects to be drawn in a realistic way but because the lines get smaller the correct measurements are not used.

Perspective drawings can have one, two or three vanishing points. Two-point perspective is most commonly used to draw products.

- A horizontal (or eye level) line is drawn and two vanishing points (VP) are drawn on.

- The vertical edge closest to the viewer is the only true measurement. The rest of the lines can be reduced in length as they recede into the distance.

- Different views of the object can be achieved by altering the view position.

    - Worm's eye view: The object is drawn above the horizon line and lets the viewer see two sides and the bottom.
    - Street level view: The object is drawn with the horizon line passing through it, two sides only can be seen.
    - Bird's eye view: The object is drawn below the horizon line and lets the viewer see the top and two sides. This is most commonly used for drawing products.

To help keep things in proportion in your perspective view a basic division method can be used.

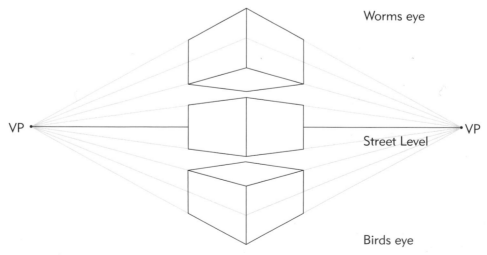

Three views in two-point perspective

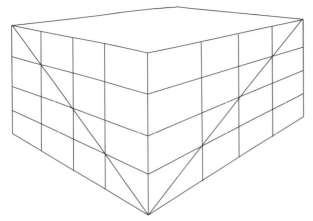

Division method used for proportioning perspective views

Isometric drawing

Isometric drawing is a formal way of showing your design. The drawing may appear distorted because it is not drawn as the eye would see the object. The correct sizes of the object are used making it an accurate method of drawing.

- All vertical lines are drawn 90° to the horizontal base line. A set square on a parallel motion is used for accuracy.

- The length and width are drawn 30° to the horizontal using a 60°/30° set square on a parallel motion for accuracy.

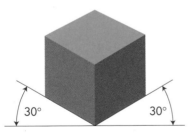

The length and width are drawn 30° to the horizontal on an isometric drawing

- All circles will appear as ellipses. An isometric template can be used for this, or the ellipse can be constructed by hand.

- All line measurements are true. They are drawn as the actual size.

- To draw curves in isometric drawings, French and flexi-curves can be used.

- Isometric grid paper can be used to save time. This paper has the 90° and 30° lines drawn on it so you do not have to use a 60°/30° set square and parallel motion.

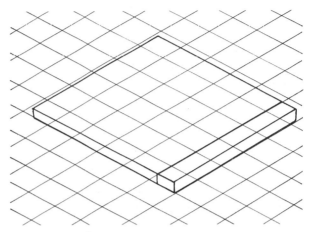

Isometric grid paper

Activity

1 Draw a cube in two-point perspective from a bird's eye view. Try to do the same from a street level view and a worm's eye view.

Key points

- Perspective drawings allow objects to be drawn as they are seen.

- Isometric drawings are an accurate method of drawing a product.

Communication of ideas – 2D views

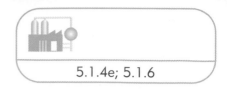

When a lot of detail is needed to be shown, it is often clearer and easier to draw it in a flat two-dimensional (2D) view. 2D drawings relate to the view in a 3D sketch and they are drawn as if you are seeing them straight on.

Orthographic drawings

This method of drawing is often used in a formal way to provide all the information needed to make the product. Designers have to produce these kinds of drawings to give to the manufacturer. This means that the drawing has to be clearly understood and must use the correct conventions. It is often difficult to visualize your product in 2D, so start by having a 3D pictorial drawing.

These three views – plan, front and end – will be drawn separately. They all relate to one another, so they are drawn on the same sheet of paper. In the third angle orthographic the plan view is drawn above the front view. The end view from the right side is drawn on the right. This allows you to relate the plan and end view to the front view.

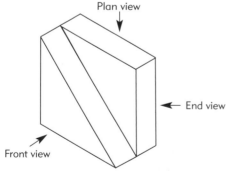

A third angle orthographic drawing

Planning and setting up the drawing

1 Get all the drawing equipment needed. You will need a set square, ruler and a drawing board with a parallel motion or tee-square. Set up your paper on the drawing board.

2 Draw a 10 mm border around the page. At the bottom leave a space for a title block. In this space you should write:

- your name
- what the dimensions are measured in
- what the product is
- what type of layout it is drawn in. Third angle is the easiest and this is what has been shown in the drawing on this page. A symbol is commonly used to denote this.

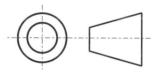

Third angle orthographic symbol

- The scale. In most cases the product will be too large to draw so it has to be scaled down. The drawing on this page is 1:2 scale (1 mm in the drawing equals 2 mm on the product), which means it is half full size. The real size of the product will have to be worked out first then it can be scaled down to fit all the elevations on the page. Try to use a 1:2, 1:5 or 1:10 scale because it will be easier to work out.

How to do it

1 Decide which is the front view (elevation) of your object. This is usually the one with the most detail on it or what naturally looks like the side of the product. Draw it in the middle of the page with enough space around it to draw the plan and end views.

The front view

2 Draw lines (projection lines) from each point on the front view vertically for the plan view and horizontally for the end view. This will save time and ensure that the sizes are all the same.

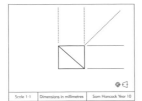

Construction lines for the plan and end views

3 Draw a 45° line from the corner of the front view.

4 Draw the end view using the construction lines from the front view as a guide.

5 Draw construction lines from the end view vertically until they touch the 45° line. Horizontal construction lines can be drawn as a guide for you to draw the plan view.

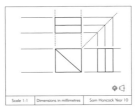

Using the 45° construction line to complete the end and plan views

Depending upon the type of product you are drawing it may be better to start with the plan view or to draw the end and plan views at the same time. This will enable you to transfer dimensions between the two easily.

Adding dimensions

When adding the dimensions or sizes to the drawing the following rules should be followed:

- Always put on the real sizes, not the scaled sizes.
- A dimension should be shown only once.
- The numbers and dimension lines should be away from the drawing.
- The numbers should be placed above and in the middle of the dimension line.
- Small filled arrows should be used.
- The radius of an arc should be marked with an arrow from the centre to the circumference.
- The diameter of a circle should be marked by double-headed arrow through the centre of the circle.

- To denote a radius, R is placed before the number and for a diameter, D is placed before the number.

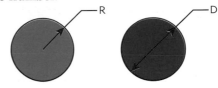

Dimensioning circles

Types of line

The lines you use to outline the product should be thicker/heavier than the rest of the lines on your drawing. The type of line you use will mean different things. In the diagram below:

1 = centre lines
2 = projection lines
3 = hidden detail line
4 = hatching/section lines.

1 — — – — – — – — – — – — – — – —

2 ————————————————————

3 – – – – – – – – – – – – – – – – –

4 ▱▱▱▱▱▱▱▱▱▱▱▱▱▱▱▱▱

Types of lines used in orthographic drawing

Activities

1 What information is included in the title block and why is it needed on the drawing?

2 Draw the symbol used for third angle orthographic projection.

3 Why are different types of line used in orthographic drawings?

Key points

- Orthographic drawings are accurate ways of drawing a product giving enough information so that it can be manufactured.
- Orthographic drawings follow common rules set out so that the viewer can understand what has been drawn.

Communication of ideas – other methods

The different methods below help to communicate ideas in more detail. You will need to select the best method for what you are trying to communicate. A number of different drawings will be easier to understand rather than trying to fit all the information on one.

Detail views

At times it can be difficult to explain details in your sketch work. Areas of the design can be sketched larger in order to offer better understanding.

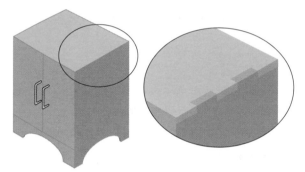

A detail view

Hidden details

Sections or parts of your design may be hidden (e.g. the tenon in a mortice and tenon joint) but may need to be shown. In order not to confuse the viewer these can be shown by drawing them with a broken or dashed line.

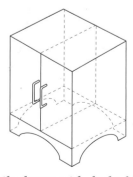

Hidden details shown with dashed lines

Cross sectional views

At times showing a lot of hidden detail can be difficult to sketch and confusing to the viewer. To simplify this you can draw the product as if a section has been cut away. To show that the section that has been cut, diagonal lines or colour can be added. To distinguish between two parts, a different colour or opposing diagonal lines can be used.

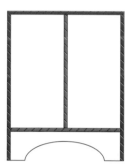

A cross sectional view

Cut away views

To show the internal views of a product, an area can be cut away to let the viewer see what is inside. This is shown by a ragged edge.

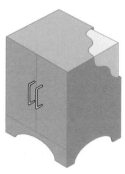

A cut away view

Exploded views

To show how parts are put together and relate to each other they can be separated into an exploded view. This allows the viewer to see all the separate parts that make up the product. When using this method the parts will have to be in the correct position so they look like they can fit together. This is often used to explain how products are assembled (e.g. with flat pack furniture). Exploded views can be drawn in both isometric and perspective. Additional lines and arrows may be used to assist the viewer in understanding the picture.

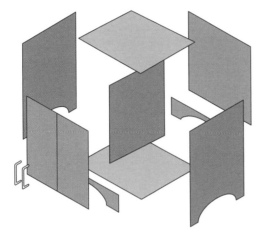

An exploded view

Annotations

Sometimes your ideas cannot be expressed through drawings or you may need to add extra detail around the drawings you have made. This can be done by adding the following information:

- Labels: To state what the view is and explain the sketch.
- Dimensions: Sizes in length, height, depth, thickness, radius, diameter etc.
- Arrows: Used to show movement and direction. They can also be used to lead the viewer's eye to the next stage.

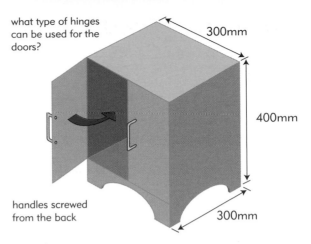

what type of hinges can be used for the doors?

300mm

400mm

300mm

handles screwed from the back

Use of annotations

In the initial ideas and development stages you may add:

- Comments: Questions, opinions and ideas that may need to be answered or addressed later.
- Evaluations: Every time you make a judgement or decision about your work, you are evaluating it. This is an on-going process and is known as formative evaluation.

Activity

1 On A4 paper draw a basic product such as a pencil sharpener in 3D. Try to explain how it is constructed using:
  - detail views
  - hidden detail views
  - exploded views
  - annotation.

Key points

- A variety of methods can be used to communicate your ideas so they are easier to understand.
- Annotation is used to help explain your drawings and record your thoughts and decisions.

Initial ideas

The essential elements you will need to help guide your design thinking are:

- your design brief
- your specification
- your research.

From this point on you can begin to produce sketches that will show a range of solutions to fulfil your brief. By sketching down your ideas you are communicating what you are thinking so you will need to present it clearly.

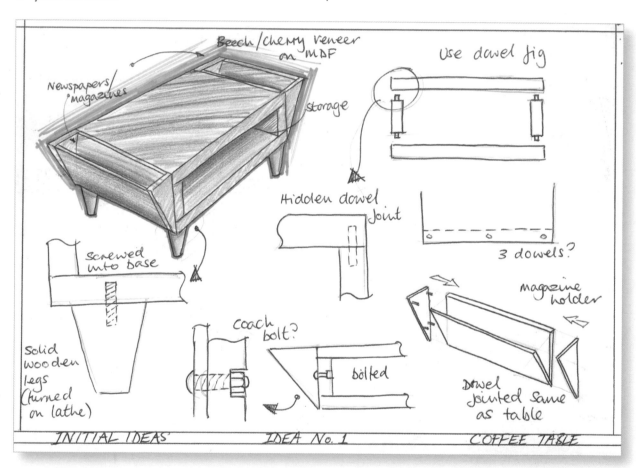

Initial ideas sheet

Let your ideas grow

Initial ideas are your very first design ideas. They are often incomplete solutions, but more detail will be added when you begin to develop the best ideas. Try to sketch down a variety of different ideas and solutions to the problem that fulfil your specification. Do not be afraid to sketch them down quickly even if they look rough, you can always add detail later to make them look better. Remember that initial ideas are not your final ideas.

Starting points

It is essential that you use your specification as a guide throughout. You can use the images you have collected in your research, such as pictures of existing products, to help start you off. The main requirement is that you should produce a range of different ideas that reflect the requirements of the specification. This means more than three, but ideally you want to produce at least six or seven ideas. For each idea you should:

- Sketch the idea down in 3D.
- Add any annotation.
- Break down the idea into details by showing how the idea might work and be constructed. Do not be too concerned with small details as this will be addressed in the development stages (if the idea gets this far).
- Add annotation.

Presentation and communication

Presentation of your ideas is important. Try to use a range of communication techniques, as mentioned on the previous pages. Your sketches can be enhanced using a variety of techniques (see *Enhancement of drawings* on p. 50). However, do not spend too much time enhancing your sketch work at this stage.

Evaluation of ideas

Once you have a range of ideas you will need to select the best ideas to develop further. You will therefore have to evaluate each idea against your specification. This can be done by producing a star diagram (see page 22) with each specification point as your criteria, and a written evaluation. In doing this you will judge which are the best ideas that can be developed further. In your written evaluation try to include:

- The good points and bad points of each design.
- Other people's opinions of the designs – these are also be useful for user feedback.
- How the design can be made commercially and in quantity.
- A concluding paragraph stating which ideas and elements you will develop.

Activities

1 What information do you need to have before you begin designing and why?

2 Why is evaluation of initial ideas important?

Key point

- Initial ideas are the first sketches you will make in order to answer your design brief and specification.

Development of ideas

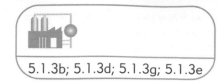
Once you have chosen the ideas you wish to develop further, use the written evaluation you have made to try to improve your designs. Developing your initial ideas will help you produce a better solution. During the development you will have to:

● Carry out extra research, specific to the problem e.g. materials that might be used, checking ergonomics, availability of parts etc.

● Evaluate your ideas against the specification throughout.

Development is not done just through sketch work. During the development of your ideas other appropriate methods will be needed, such as:

● Producing models to test and refine ideas (see page 50).

● Using ICT applications to test and manipulate designs (see pages 54–59).

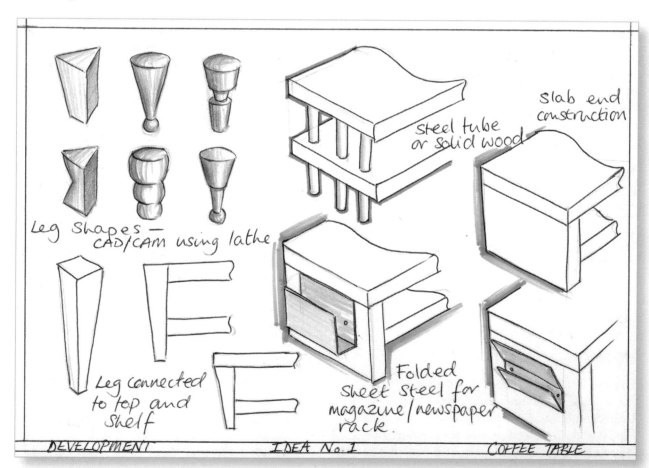

Development sheet

Start by sketching your initial idea. You will have to ask yourself a series of questions and use different communication techniques to answer them. In doing so you will show your progression of thought in making the idea evolve. There are three important areas to keep in mind while developing your ideas.

- Function: How will it work?

- Aesthetics: How will it look?

- Construction: How will it be made?

With these in mind you should answer the following questions during the development of your product:

- Alternatives: Are there different ways I could do this?

- Improvements: How can I do this better?

- Refining and simplifying: Is there an easier way?

Do not try to develop an idea all at once. Break it down into smaller sections and develop these; for example, if you are developing a coffee table you could break it down into the following sections:

- The shape of the top.

- The shape of the legs.

- How to attach the legs to the top.

- The materials you could use.

- The size.

- Extra features you could include, such as a draw or a magazine holder.

Evaluation

Throughout the development of your idea you should:

- Produce sketches, formal drawings, models and ICT work.

- Evaluate your ideas using annotations and reports, justifying decisions you have made and comparing your ideas against your specification.

At the end of the development you will have a range of different ideas and possibilities. You will now have to make a decision about which is the best solution. Write down why you have chosen it and any further development that is needed.

The chosen idea

The chosen idea will need some further development. Continue to develop the design to the point where you have worked out all of the problems (and potential problems). It is better to do this now so that when you come to planning and making the product you will have a much clearer idea of what to do.

Activities

1 What three areas are important to keep in mind when developing your ideas and why?

2 Why is evaluation so important?

Key points

- Development is the evolution of your ideas by showing improvements, solutions and possibilities.

- Development can be carried out through drawing, ICT and model making.

Modelling

Modelling ideas in development is essential. It is sometimes hard to work out how an idea will work or see how the design will look only by sketching. Models can be made to help you and there are different models that can be used at different stages of the design process.

You may need to scale your model up or down (1:2 = half size; 1:4 = quarter size). This will save time and material while still giving an accurate representation of the product.

Sketch model

This is the first model you will produce in order to expand on your sketched ideas. It will help explore the shape and proportion of your possible ideas, as well as working out any problems.

A range of sketch models should be used as part of your development. It is often easier to see areas that need developing when your idea is in 3D. This allows for in-depth development to occur.

Materials such as card and Styrofoam are commonly used for sketch modelling. They are inexpensive, little equipment is needed and, more importantly, they can be worked quickly. The surface finish is unimportant as the emphasis is on the development of the design.

Models that are produced should always be kept, as they may prove useful in later stages. They also show your design thinking and form an important part of your folder work. Therefore, try to photograph your models so that you have a permanent record in your sketch work.

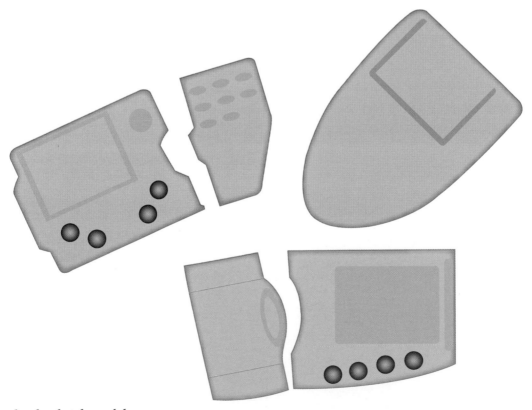

An example of a sketch model

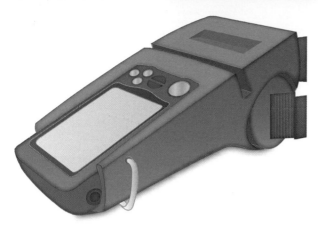

An example of a block model

Block model

This is an accurate representation of the design and will determine the shape, dimensions and surface finish. Your model should show as much surface detail as possible (screws, joints, texture, colour and finish) but have no internal details of moving or working parts. This type of model is produced once the working drawings have been finalized.

Working model

This is the same as a block model but includes basic internal detail showing moving and working parts.

Prototype

This is the final complete working product that you will produce after design and planning. A prototype is usually made by hand as a one off and is made from the same materials that are to be used in the final product. This allows for evaluation and adjustment before money is invested in tooling for large scale manufacture.

Modelling materials

Material	Sketch Model	Block Model
Paper	X	
Card	X	
Foam board	X	
Papier mache	X	
Expanded polystyrene	X	
Corriflute	X	
Art straws	X	
Wire	X	
Balsa Wood	X	X
Styrofoam	X	X
Jelutong		X
Metals		X
Woods		X
Plastics		X

Activities

1 Why is scale used?
2 At what stage of the design process would you use:
   ● a sketch model?
   ● a block model?
   ● a prototype?

Key point

● Models are used in the design stages to help the designer visualize and develop designs.

Enhancement of drawings

A variety of techniques can be used to enhance and improve your drawings. This can be done so that your drawings:

- look more attractive
- are highlighted (or parts of them are) and brought to the attention of the viewer
- appear more realistic.

Media

A range of different media can be used for enhancing your design work such as:

- pencil crayons
- studio markers
- paints
- pastels
- pens.

Studio markers and coloured pencil crayons are the quickest to use. They come in a wide range of colours and they do not have to be mixed or left to dry. Studio markers are excellent for large blocks of flat colour but it is difficult to achieve gradual tone. Pencil crayons can be used on top of markers to give tone or they can be used on their own. Before applying any media to your design work test it out on scrap paper.

Tonal shading

All objects have light cast on them either from a natural source (daylight) or an artificial source (light bulb). The object will therefore have lighter and darker areas depending where the light source is. If you apply light and dark shading to you drawing it will make it look realistic, as it would appear to the viewer in real life.

When applying tonal shading to a drawing you will need to decide where the light source is (it is usually over your left shoulder). The areas

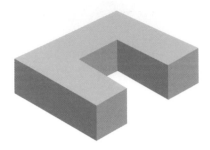

Tonal shading

closest to the light will appear the lightest (light tone) and the areas furthest away will appear the darkest (dark tone). The areas that receive some light will be between the two (mid tone). Shadow can then be added.

Materials

In order to communicate what materials are to be used to make your product you will need to render your drawing to try and represent the material being used. The obvious way of doing this is by using colour, but with materials such as plastic that come in a variety of colours this would be hard to do. Therefore you should try to represent the nature of the material's surface, for example it may have a grainy, reflective, shiny or transparent texture.

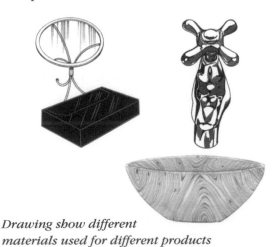

Drawing show different materials used for different products

Background and outline

This technique is used to make a drawing, or parts of a drawing, stand out from the page and add extra information.

● Outline: The drawing can be outlined with an ink pen. Colour can also be added around the drawing.

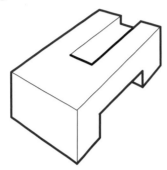

Outlining with ink

● Frame: A block of colour or images that relate to the product can be drawn behind the sketch.

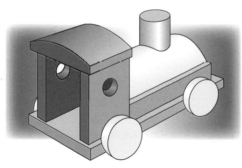

Framing a sketch

● User outline: By drawing the outline of the user on your sketch it will give the viewer a sense of scale. Having the user interacting with the product will also add to the viewer's understanding of the function of the product.

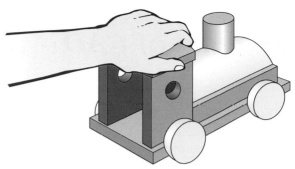

Outlining the user in a sketch

● Environment: In the background you may insert an image of where the product will function. This will give a better idea of scale as well as how it will fit into its surroundings. The easiest way of doing this is by placing a scanned image or photocopy behind the drawing.

Presentation drawings

As well as producing a block or working model to visualize your design, a presentation drawing is also useful. Designers use this type of drawing to present ideas to clients or manufacturers. The design is often drawn using a formal 3D method so it looks as realistic as possible. This means that tone, material and background techniques are all used.

Activity

1 Sketch three cubes.

   a On the first add tone, an ink outline and frame it.

   b On the second draw it as if it were wood and outline it in colour.

   c On the third add tone using one colour, and use an image from a magazine that relates to it as a background.

Key point

● Different methods and techniques can be used to improve and add more information to a drawing.

The use of CAD

CAD (Computer Aided Design) can be used to model ideas in 3D as well as produce final accurate drawings in 2D. Most CAD packages available are 3D; this is where the product is drawn in 2D and 3D to make a solid shape. The shape can then be revolved to allow the user to view the object from any angle.

Most CAD packages will allow the user to assign material to the object. For example, a table could be viewed in metal, wood or plastic. You can also control the lighting and camera angles to create a realistic image. It is therefore possible to create 'virtual' products and environments on CAD software. Products can also be animated to move on screen to give the viewer an idea of how the product may work.

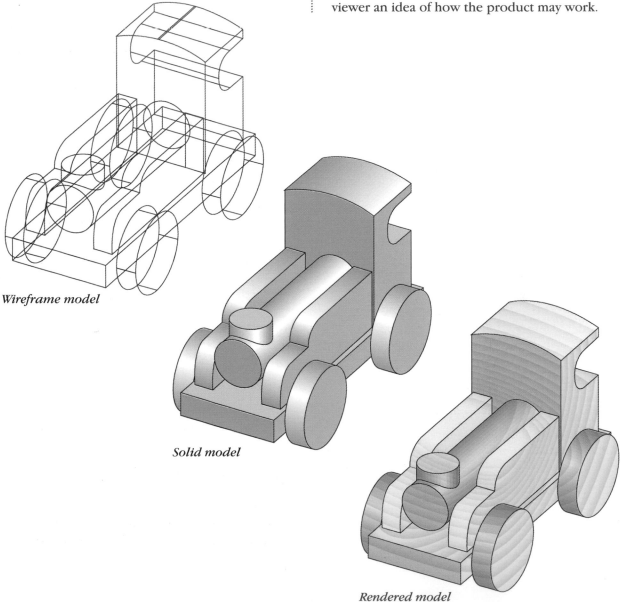

Wireframe model

Solid model

Rendered model

Modelling with CAD

There are three types of model you can produce using CAD software: a wireframe model, a solid model and a rendered model.

Wire frame model

The 3D object is represented by a series of lines. These lines can be removed and the drawing enhanced to make it appear solid.

Solid model

The 3D object appears solid. This can follow on from wire frame models.

Rendered model

The surface of the 3D object is assigned a colour or materials (such as glass or wood) to make the object appear realistic. Lighting can be altered to enhance the appearance and give a stronger 3D form.

The advantages of CAD modelling

- Designs can be electronically shared with others so that more than one person can work on a design at once. This allows fast development of ideas whether the designers are in the same office or on the other side of the world. Therefore the designs can be tested, evaluated and modified quickly without the need for costly prototypes and models.

- Information about the product can be saved and stored for when it is required.

- The design can be manipulated without losing key information.

- Photo realistic models can be produced and animation applied to show a virtual prototype.

- Real 3D models can be created using the data from the CAD package. This allows for any potential errors and problems in tooling and production to be detected early on in the design process. It also allows the user to physically interact with the product, gaining a better understanding of its size and shape. Rapid prototypes are accurately produced in layers of resin. The data from the CAD package is converted to control the CNC (computer numerical control) machinery. This controls lasers which solidify liquid polymers in resin. This production is a one-off which saves time, money and complex tooling.

- Rapid prototyping (RPT) is an expensive way of producing the product from CAD. A variety of CNC machinery can be used to produce block models (see pp. 56–7). Styrofoam is often used, as it is inexpensive and quick to work with.

The disadvantages of CAD modelling

- The hardware and software needed to carry out modelling can be expensive.

- Training is required to use CAD software efficiently.

- CAD modelling can be time consuming, especially if you want to produce quick sketch models as part of development.

Activities

1 Why are rapid prototypes produced?

2 Why is it important for modifications to take place?

3 How does the manufacturer save money?

Key point

- CAD is used to speed up the design process. As a result, a greater variety of designs can be produced and evaluated, with modifications added quickly.

CAD/CAM in industry

CAM (Computer Aided Manufacture) is a term used to describe the manufacture of products or parts by machines that are controlled by a computer. CAM has been a major contributing factor to the improved quality of mass produced goods as well as lowering the production costs. Some manufacturers now have fully automated production, and the only people employed are there to monitor the production.

CNC machinery

The machines that manufacture products can range from robotic arms to engravers. They can be individually programmed and most are computer controlled. One of the most widely used type of machine is CNC (computer numerical control) that provides a link between CAD and CAM.

The data in the CAD program can be transferred into a machine control unit (MCU), which converts the information in order to control the measurements the CNC machine will make.

Types of CNC machines available are:

- lathe
- milling machine
- engraver
- router
- drilling and cutting
- pressing and punching
- vinyl cutters.

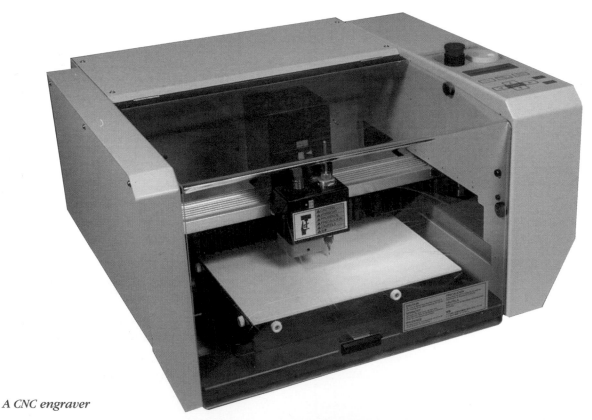

A CNC engraver

A milling machine

Advantages of CAM

- Machines can operate in environments that might be harmful to humans.

- New instructions can be given by changing computer programs, thus giving the manufacturer more flexibility.

- Quicker production times. Complex shapes and designs can be created easily.

- Accuracy and quality in the production machines can be controlled with few errors and they are not affected by human constraints.

- Repeated production.

Disadvantages of CAM

- Costs of buying and installing machinery are very expensive.

- The loss of jobs for workers.

- Workers at fully automated factories have very little job satisfaction, as all they have to do is monitor the machinery.

Activities

1 Why is CAD/CAM used?

2 Why are workers' jobs lost because of it?

Key point

- CAD/CAM has resulted in the manufacture of better products at a lower price.

Using ICT

ICT stands for Information and Communication Technology. As well as CAD/CAM, ICT can be used to help you with your folder work. As part of your GCSE coursework the use of ICT should be included.

ICT can be used throughout your design folder. It will help to improve the quality of presentation, save you time as well as offer useful tools such as the manipulation of images and text.

The areas ICT can be used in are:

- research
- the presentation of folder work
- communication of information
- development of ideas.

Research

The use of the Internet allows you to access information easily and quickly. You will be able to visit a company's site to find out about existing products or visit virtual factories to see manufacturing taking place. You can also e-mail companies about their products and ask experts for advice.

Presentation of folder work

DTP (desktop publishing) packages such as 'Publisher' and 'Quark' allow you to manipulate text in typeface, style, size and colour. The overall page layout, including images and pictures can also be manipulated. This process has been used to produce such items as adverts, magazines and the book you are reading now.

Communication of information

ICT packages such as spreadsheets give you the option to pictorially show information in the form of graphs, charts and tables. These can be used to show results from questionnaires you have carried out. Showing you results pictorially allows them to be easily understood.

Tips

- When researching information from the Internet, you will often use a search engine such as 'Google'. Try to be specific when typing in key words otherwise you will spend a lot of time searching through hundreds of irrelevant web sites. For example, if you want to find out about coffee tables and side tables on the market, don't type in 'table'. This would result in the listing of hundreds of web sites relating to all kinds of tables – results tables, football tables, periodic tables, etc. Be more specific by using phrases such as 'furniture + table' or 'furniture + retail'.

- Visiting design and technology-related web sites can be a good starting point. These may offer suggestions of where to visit, or even provide links to other sites.

- Most web sites have links to other related sites. These can be useful in finding information specific to your area of research.

Development of ideas

Images and pictures can easily be manipulated on the computer. Image manipulation software packages, such as 'Paint Shop Pro' and 'Photo Shop' are often used to do this. The image first has to be captured using a scanner or digital camera. It is then converted to a file so it can be opened up in the image manipulation package and saved.

Sketches and formal drawings can be scanned and the design can then be manipulated through the application of colour, texture and text. Backgrounds and outlines of users may then be added to help present your ideas.

Models you have produced can be placed in an environment or situation. This adds to the viewers' understanding of the products' function and use.

Activity

1 What does the following software allow you to do in your folder:
- DTP?
- Paint Shop Pro?
- Spreadsheets?

Key point

- The use of ICT in the folder work can enhance presentation and communicate ideas. It will also allow you quick and easy access to information.

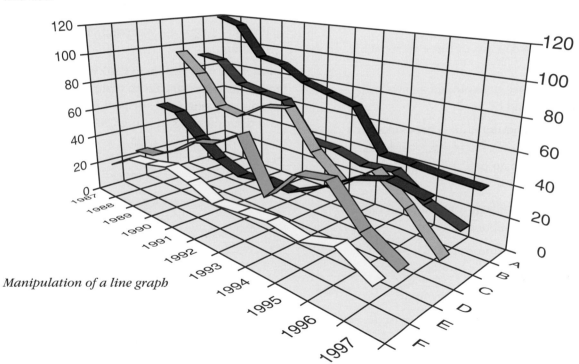

Manipulation of a line graph

Questions

1 When designing, it is important to communicate your ideas in a clear way.

   a What is informal drawing and when is it used?

   b What is formal drawing and when is it used?

   c Name three different types of specialist drawing equipment used in formal drawing.

2 There is a wide variety of methods that can be used to communicate your ideas.

   a Name two types of 3D pictorial drawing commonly used by designers.

   b 2D drawing consists of three views. What are they?

   c Describe what an exploded view is and when it would be used.

   d Explain why annotation is used on sketch work.

3 Examine the following design brief and specification:

   Design brief: To design a kitchen roll holder to be used at barbeques.

   Specification: The product must –

   ● be free standing

   ● be one metre above the ground

   ● allow the user to tear off a piece of the roll using one hand

   ● allow the kitchen roll to be easily replaced.

   a Using sketches and annotation produce three initial ideas.

   b Choose your best design and develop it using a variety of communication methods.

   c Present your chosen design as 3D drawing with tonal shading.

4 Before your design is manufactured, a prototype would be made.

   a Explain the term 'prototype'.

   b Why would a prototype be made before it is produced in quantity?

5 Your design could also be modelled on a CAD package. List three advantages of CAD modelling.

6 Modelling your design in CAD will allow data to be easily transferred so that parts can be produced using CAM. Discuss the advantages and disadvantages of using CAM.

PRODUCT DEVELOPMENT

Metals

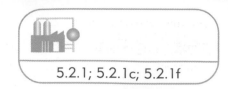
Metal is an extremely versatile material used to manufacture an extensive variety of products. There are a range of metals available and each has its own distinctive properties. When selecting a metal to use it is important to choose the most appropriate one for the task at hand.

Advantages and disadvantages of using metals

Metals have the following advantages:

- They can be treated and combined to create useful properties.
- They are malleable and ductile.
- They are strong, tough and durable.
- They are good conductors.
- They are recycled easily.

Some disadvantages in using metal include:

- Some metals can corrode easily.
- They are heavy.
- A lot of energy is used in extracting a metal from its ore.
- Pollution is created during the extraction of metals.

Classification of metals

All metals fall into one of two categories:

- Ferrous metals: Metals which mainly contain iron or ferrite. These are all magnetic.
- Non-ferrous metals: Metals which contain no iron or ferrite. These are not magnetic.

Both of the above can be sub-divided into two categories:

- Pure metals: Metals which are made of one single element.
- Alloys: Metals which are made by combining a number of pure metals, and sometimes other elements, to create useful properties. For example, stainless steel is a ferrous alloy. It has chromium and nickel mixed into it so it will not corrode or rust. It can be used for making sinks and kitchen utensils because it is hard wearing and will not rust, making it hygienic to use.

Metal forms

Metals are produced in a variety of forms:

- sheet
- tube
- box tube
- bar (threaded bar)
- flat bar
- angled bar
- hexagonal bar
- extruded form.

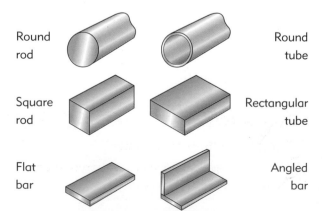

Some commonly available forms of metal

Common Ferrous Metals

Name	Composition	Properties	Uses
Mild steel	Alloy (Iron and Carbon)	High tensile strength, ductile, tough and fairly malleable.	General purpose steel – sheet, tubes, girders, nuts, bolts etc .
High Carbon Steel	Alloy (Iron and carbon)	Hard, tough and malleable. Harness and toughness can be improved by heat treatment.	Hammers, chisels, drills, files, lathe tools.
Stainless steel	Alloy	A hard, tough and malleable metal which is corrosion resistant.	Sink units, kitchenware, medical equipment, pipes, aircraft.
High speed steel (HSS)	Alloy	Retains hardness at high temperatures but can also be brittle.	Drills, machine tools, cutting tools for lathes.

Common Non-ferrous Metals

Name	Composition	Properties	Uses
Aluminium	Pure	A light, soft, malleable, ductile material that is highly conductive to heat and electricity. It has a relatively low melting point and can be soldered and welded by special processes. Aluminium is also corrosion resistant.	Aircraft, bikes, cars, window frames, castings, kitchen equipment, packaging, insulation, cables.
Copper	Pure	Malleable, ductile and tough which is highly conductive to heat and electricity. It is corrosion resistant and can be easily soldered.	Wire, cables, pipes, printed circuits, saucepans, roofing.
Brass	Alloy	A hard material that is corrosion resistant but tarnishes easily. It is a good conductor and can be cast, machined and soldered easily.	Castings, ornaments, tools, forgings, valves.
Bronze	Alloy	Harder and tougher than brass, it is a hardwearing and corrosion resistant material which can be easily machined.	Valves, bearing, springs, castings, gears.
Tin	Pure	A soft material, with a low melting point, which is corrosion resistant in damp conditions.	Often used to make tinplate, a thin sheet of steel coated with pure tin. It protects the steel from corrosion and is non-toxic so it is used for food cans.

Activities

1. What benefits do alloys offer?
2. What impact does the production of metal have on the environment?

Key points

- All metals can be classified into categories.
- There is a wide variety of metals, each with distinct properties.
- It is important to select the correct type of metal to use for a product so it will be fit for its purpose.

Wasting metals

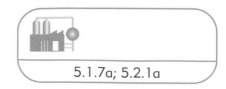
Wasting is a process whereby metal is removed. When removing metal through cutting and filing, burrs are created. These are fine shards of metal and for this reason you must be careful when handling metal. Ensure the burrs are filed off by draw filing at a slight angle. Low gauge metal can also be very sharp. Ensure that gloves are worn when handling this type of material.

Filing

A file cuts on the forward stroke and takes off small filings. The amount of material removed will depend on the type and grade of the teeth on the file. The appropriate file must be used for the task at hand. Files come in a variety of shapes, sizes and cutting profiles. When selecting a file you will have to consider how much material you want to remove and the shape of the material. As a general guide:

- for flat surfaces, edges and external curves flat files are used

- for internal curves, round and curved files are used.

Types of file

Files are graded and range from smooth (100 teeth per 25 mm) to rough (10 teeth per 25 mm).

- A single cut file is used for soft materials and has a smooth finish. It has an 80° angle and will not clog.

- A double cut file is used for medium to hard materials and has a general purpose finish. One row of teeth is set at an 80° angle and the second row is at a 60° angle.

- A curved tooth file is used for the rapid removal of soft materials without clogging.

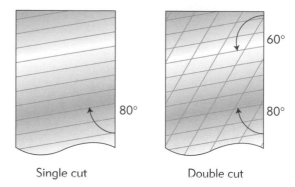

Single cut and double cut files

Filing methods

- Cross filing: This method of filing ensures rapid removal of material but does not leave a smooth finish. The full length of the file is moved across the material at an angle.

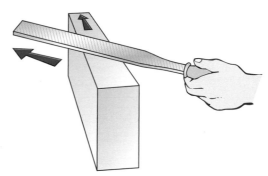

Cross filing

- Draw filing: This method does not remove material quickly but does allow a smooth finish to be obtained. The file is moved sideways across the work and at 90°.

Draw filing

Cutting

For cutting tubes, bars and plates, a hacksaw is commonly used. Smaller hacksaws are known as junior hacksaws. Hacksaw blades are held within a frame and can be changed according to the material to be cut. The cut is made on the forward stroke so the teeth face forwards. The depth of the cut is limited to the depth of the frame.

A hacksaw

Hacksaw blades are graded by the number of teeth:

- For cutting material up to 3 mm thick use a blade with 32 teeth per 25 mm.
- For cutting material 3–6 mm thick use a blade with 24 teeth per 25 mm.
- For cutting material over 6 mm thick use a blade with 18 teeth per 25 mm.

For cutting low gauge material, snips are commonly used. These can be held or placed in a metal vice.

- Hold the snips with the thumb and fingers around the handle but do not put any fingers between the handle as they could get trapped.
- Push the snips into the work to open them, and as you are cutting do not close the snips completely (this is only done at the end of the cut). To cut curves and complex shapes remove the majority of the waste and leave 5–10 mm around the edge you want to cut. You can then cut the line with greater accuracy.

Snips

For rapid shaping of a thin sheet, an electric jigsaw can be used, fitted with the appropriate blade. To prevent the sheet from vibrating and causing a damaged edge it will have to be sandwiched between material such as 4 mm MDF (medium density fibreboard) before cutting.

Activity

1 Describe what methods and tools you would use to remove 3 mm of low gauge mild steel from an internal curve and achieve a smooth finish.

Key point

- There are two ways of removing material by hand: filing and cutting.

Machining metals

Machining is a common term used to describe how metal is accurately shaped, through the removal of material on a lathe. A lathe can rotate cylinders of metal. Centre lathes allow the accurate production of cylindrical, conical and curved shapes from a solid rod. They also allow holes to be bored and flat surfaces obtained across the ends of rods

The lathe needs to be set up correctly for use. The cutting speed will depend on the material being cut and the process undertaken. Cutting fluids may have to be used to cool and lubricate the material being cut. Most material will be held in a three-jaw chuck, but for longer rods a support called a centre will also have to be used at the other end.

There are a variety of cutting tools that can be used depending on the process. A right hand knife (cuts left to right) or round nose (cuts both ways) are the most common. The lathe tool should be set up in the centre of the work. The methods used in machining are described below.

Facing off

This technique is used to achieve a flat surface at the end of a cylindrical rod. The cutting tool has to be set up at 90° to the centre of rotation. By moving across the surface of the rod from the outside to the centre a flat surface is produced.

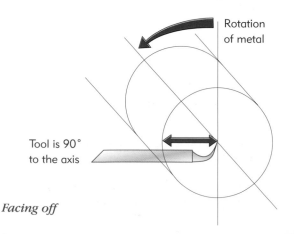

Rotation of metal

Tool is 90° to the axis

Facing off

Parallel turning

This technique is used to remove material along the length of the work. This can be used to reduce the diameter of the cylindrical shape. The cutting tool moves sideways, parallel to the axis of rotation. This technique can also be used to produce conical shapes or tapers. This can be achieved by adjusting the cutting tool so it cuts at an angle to the centre of rotation.

Drilling

Accurate holes can be drilled into the ends of the material by mounting a drill into the tailstock and gradually moving it into the material. It is not possible to centre punch the material before drilling, so a centre drill is used first. A centre drill is used on a lathe to locate the drill by making a small hole first.

Milling

Milling is a process that uses a multi-toothed cutter to remove metal from a solid block. Milling can be carried out either vertically or horizontally. In either case the work (metal block) is clamped firmly to a table below the cutter. The table can then be moved in three different directions:

- x axis – along the length of the table
- y axis – across the width of the table
- z axis – vertically up and down.

The milling process can be used to produce flat horizontal and vertical surfaces. Slots and holes can also be machined into the work. Milling is an extremely accurate method of shaping metal.

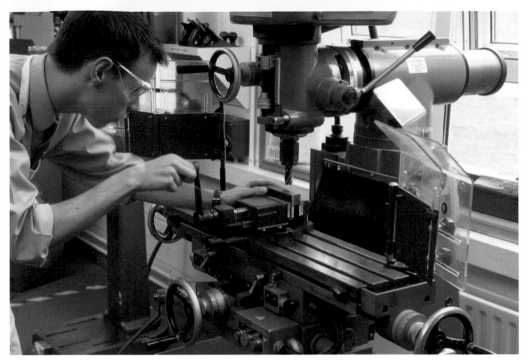

Milling

CNC machines

CNC milling machines and lathes are often used in industry. They allow for repeated production while maintaining accuracy. The labour costs are confined to setting up the CAD/CAM program and maintaining the machine while it is producing the product or batches of products. This will save production time as well as avoiding human error.

Using CNC machines in production also allows for flexibility. A variety of different CAD/CAM programs can be used in the production of different products. A batch of products can be produced using a particular CAD/CAM program. The program can then be changed to produce a batch of different products. This flexibility allows the manufacturer to avoid waste and keep up with market demands.

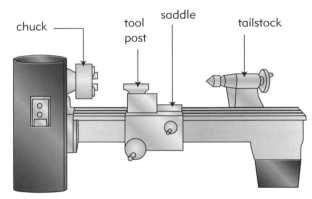

Metal turning lathe

Activities

1 Explain why the following are used:
   - different cutting speeds
   - cutting fluids.

2 Name a product that has been machined.

Key point

● Machining is a highly accurate way of shaping cylindrical metal rods.

Deforming and reforming metals

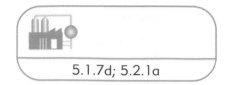
The shape of a metal rod can changed by deforming, and reforming is a process where shapes can be made by pouring molten metal into a mould.

Deforming

Deforming is a way of changing the shape of a metal. Existing forms that are produced (rod, bar, sheet etc.) can be re-shaped to produce a desired form. The methods used in deforming are described below.

Forging

This requires that the metal is hot so it can be hammered, twisted or pressed into the required shape. Forging is often used to produce strong tough shapes from black mild steel and tool steel. The hammering the materials receive helps refine the grain in the metal and increases the toughness.

Metals such as copper, brass and silver can be shaped while cold because of their malleable nature. This is known as beaten metal work, and it is often used for bowls, jugs and vases.

Pressing

Pressing cold thin sheets of metal is often used in mass production to produce shapes such as car body panels. A male die and female mould are produced to the shape required and the blank metal sheet is placed in. The male die applies pressure and forms the shape in the female mould.

Folding and bending

Thin sheets of metal can be easily bent after annealing (softening) by heat. A former or jig can be made if several items need to be shaped or a specific diameter of bend is required.

Bending sheet material can be done by using folding bars to give a straight clean fold. Folding bars are placed along the fold line and placed in a clamp. The metal sheet can then be folded using a soft faced hammer or mallet to prevent damage to the metal. Two piece folding bars with handles give a quicker and neater finish. In industry, automated and hand operated folding machines are often used.

Metal being folded using folding bars

Bending tube is more difficult as both sides of the tube need to be supported. This can be done using tube bending machines.

Reforming

Reforming, or casting, is a process where shapes can be made by pouring molten metal (metal which has been melted by heat into a near liquid state) into a mould. When the metal has cooled it returns to a solid state. This allows for products and parts to be made which would be difficult to produce using other methods. Aluminium is often used in this process because it has a low melting point and is a soft metal so it can be easily finished.

Sand casting

This is commonly used in industry to produce small production runs and large and intricate shapes. A mould is produced by making a pattern (often from wood) and compacting sand around it, mixed with silica and binder. When the pattern is removed there will be a hollow in which the molten metal can be poured.

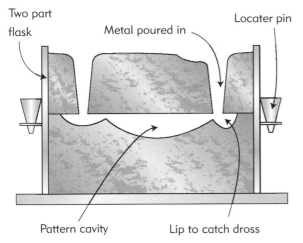

Sand casting

The mould is often made in a flask that has two parts that locate accurately together but can be easily separated. The mould must be made with a hole to allow the metal to be poured in. The casting will not have a high quality surface finish so it will need to be filed and polished later. It can be improved by using oil-bound sand or a mould coating.

Lost wax casting

This is often used for intricate one-off products where the pattern is expendable. The pattern is made from wax and is set in sand. When the mould is heated the wax runs out so that the molten metal can be poured in. This can be done with expandable polystyrene as it vaporizes when the metal is poured in (this must be done in an extremely well ventilated area).

Heat treatment of metals

Heat treatment is a way altering the properties of metal and making the metal more suitable for its purpose. The three basic stages involved are:

- heating the metal to the correct temperature;
- keeping it at the correct temperature for the required amount of time; and
- cooling the metal in the correct way.

For example, to make a chisel from carbon steel it will have to be annealed (softened) so it can be shaped. It will then have to be hardened, to increase hardness and tensile strength. Finally it will have to be tempered, a process which removes the brittleness from the metal and makes it tougher so it can be used.

Activity

1 What are the advantages and disadvantages of:

- deforming?
- reforming?

Key points

- Deforming is a process by which existing metal forms can be reshaped.
- Reforming is a process by which metal is heated, melted and reformed to a shape using a mould.

Fabrication of metals

Fabrication is a process where metals can be joined. Metals can be joined in permanent or non-permanent ways. You therefore have to be careful in selecting the most appropriate method.

Permanent fabrication

Soldering and brazing

Sand brazing work involves heating two pieces of metal to be joined, along with a filler rod. This filler rod is made of an alloy with a lower melting point than the metal. When the filler rod melts it forms an alloy with the base metals and cools to join the metal as 'glue'. The higher the temperature needed, the stronger the joint will be.

A flux (which comes in powder or paste form) has to be used with the filler rod in order to protect the surface from oxidization when it is heated. This is so that the joint will remain clean for the filler rod and metal to stick together. The flux is also necessary to allow the melted filler rod to run into the joint.

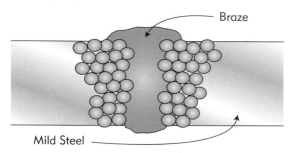

Brazing

- Clean the joint with emery cloth and make sure it fits together well.

- Mix the flux together with water so it is creamy and paint it into the joint.

- Heat the joint gently with a blow torch in order to dry the flux, and then continue to heat.

- Heat the end of the filler rod and dip it into the flux so the flux sticks to it.

- Touch the end of the filler rod onto the joint. If the join is hot enough the filler rod will melt and run along the joint.

- Allow the joint to cool.

Welding

When two pieces of the same metal are welded they are heated to the point where they melt together. On cooling they solidify and fuse to form a joint that is as strong as the metal around it. The filler rod is melted at the same time into the joint to fill in the gap. The filler rod is made from the same metal as the metal being welded.

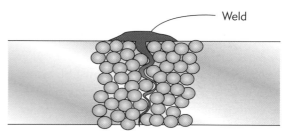

Welding

There are three common types of welding:

- Oxy-acetylene welding: Gas welding using a blowtorch.

- Arc welding: Welding using an electrical current to fuse the metal together. A filler rod is attached to a handle with the current running to it. Another clip is placed on the work, and when the rod touches the work the electrical circuit is complete.

- Mig welding: Welding using an electrical current to fuse the metal together. This is much the same as arc welding, but continuous filler wire is fed to the work from a spool.

Non-permanent fabrication

Pop-rivets

Pop rivets are used to join sheet metal or tubes together. Holes are drilled through both pieces to allow a rivet to pass through with clearance at the back. A pop riveting gun is used to pull the mandrel expanding the rivet head. When enough pressure has been applied, the tail of the mandrel breaks off leaving the head in the rivet.

Nuts, bolts and screws

These fixings come in a range of shapes and sizes. They are usually made from steel or brass. Steel fixings can be coated to prevent rust or improve their appearance. They have the added advantage of allowing the parts to be joined and disassembled easily.

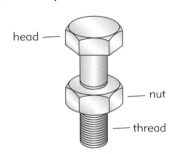

head — nut — thread

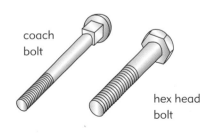

coach bolt | hex head bolt

nylon locking nut | wing nut | standard hex nut

Taps and dies

A split die and die stock is a handheld tool which can cut external screw threads on a metal rod. Different size split dies can be used, dependent on the diameter of the rod. It is essential to place the die square to the rod. If it is not square, then the thread will become 'drunken'. Once the die is on it is turned clockwise to cut the thread, then turned anticlockwise to clear the swarf. Lubricant is used to help cut the thread.

Taps are used to cut internal threads. A hole is drilled using a drill that matches the tap size. This should also match the external thread of the bolt. The tap is held in a wrench and turned clockwise to cut the thread.

Activities

1 Why is a flux needed in the brazing process?

2 Why does welding give a stronger joint than brazing?

3 What advantages does non-permanent fabrication have?

Key point

● Metals can be joined together in a variety of ways. This will depend on what the product is being used for and what needs to be joined.

Finishing metals

Metals can be finished in a variety of ways. How a metal is finished will be dependant on:

- What its function is, for example protection from corrosion or durability in use.
- The desired aesthetic qualities such as colour and texture.
- The type of metal used.

Metals can be finished using applied finishes such as paint/lacquer/dip coating or using self finishes such as polish.

Preparation

To ensure a good finish the final product will have to be cleaned up and prepared. The three basic types of preparation are:

- Grinding/filing: A revolving disc on a grinder can remove material quickly and it can also cut the metal if required and remove uneven welds, splatters and burrs. A file can also remove welds, splatters and burrs. Work the file in the same direction along the metal and finish by draw filing.
- Rubbing with emery cloth: Using different grades of emery cloth; rough, medium and smooth, marks can be removed easily. Water or oil can be used on fine grade cloth or silicone carbide paper to ensure a good finish.
- Pickling: Most metals, but primarily copper and brass, can be cleaned using pickle. This liquid is one part sulphuric acid and ten parts water, and corrodes the metal when it is placed in it. After the metal is removed it should be washed to stop any further corrosion.

Types of finishes

Before the application of paint or lacquer, ensure that the product is clean and degreased. This can be done by using hot water and detergent or a special degreaser. As painting can be a messy process, prepare the area for painting and support the work while painting and drying.

Paint/lacquer

With paint, you will need to apply primer, undercoat and a topcoat using a clean dry brush. Always paint in one direction and lightly rub down in between coats with a fine grade paper. To produce a flatter, even coat, spray paint can be used. Ensure this is done with the correct ventilation and equipment. When applying lacquer, the surface of the metal must be prepared well. Lacquer is clear, so any marks left will be seen through it. Apply at least two coats to give a durable finish.

Dip coating

This method is widely used in industrial processes that coat metal with plastic. It is used to protect metal from corrosion and ensure a durable finish.

- Heat the finished object.
- Suspend the object in a fluidization tank. In this tank thermoplastic-polythene is commonly used and has air blown through it. In doing so the tiny plastic particles behave like a liquid that allows the object to be evenly coated.
- Re-heat the object to fuse the plastic, then hang it up to cool.

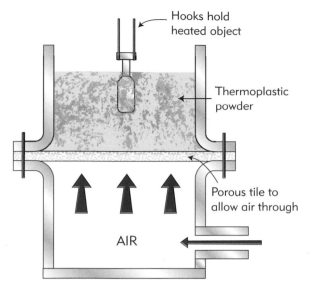

Dip coating

Metals can also be electroplated. This is where a thin layer of metal is electrically fused on to the metal's surface. Chromium (chrome) is commonly used in electroplating.

Polishing

Metals such as stainless steel, brass and copper are often polished to enhance their appearance and properties. Once the metal has been cleaned and all the scratch marks have been removed it can be polished. Buffing machines and polish are used to achieve the desired finish.

Activity

1 For each type of finish give an example of a product which has that finish. Explain why it has that particular finish.

Key point

● Applied finishes offer both functional and aesthetic qualities.

Smart materials

'Smart materials' is a term to describe modern materials. These materials have been designed to provide the best properties and behaviour possible for their intended uses. As manufacturers experiment with materials, they are discovering ways the materials can be adjusted and altered. Smart materials, particularly the following, are becoming more readily available:

Polymorph

A thermoplastic that becomes mouldable in water at 62 degrees.

Smart memory alloys (SMA)

A metal that changes shape at a set temperature.

Thermochromatic pigments

These can be added to paint or materials and change colour when they reach a certain temperature.

Smart grease

A lubricant that can be added to mechanisms to regulate speed and thus produce a smoother motion.

Softwoods and hardwoods

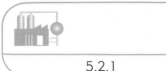

Different species of trees produce different types of wood. The properties of the timber will depend on what type of tree the wood comes from. Once the tree has been felled it will be taken to a sawmill and dried out to prevent it from shrinking and warping. It is then converted into workable planks of wood.

Deforestation and sustainable wood

The use of timber has become a major environmental concern. Large areas of forest have been cut down resulting in long term effects to wildlife and indigenous populations across the world. This deforestation often takes place in warmer climates that produce slow growing hardwood trees. Therefore it would take hundreds of years for these trees to grow back.

As a result some countries grow trees as a crop to harvest. This sustainable resource often tends to be quick growing species.

Types of wood

There are three types of wood available:

- hardwood
- softwood
- manufactured board (see page 76).

Common Hardwoods

Name	Advantages	Disadvantages	Uses
Beech	Hard, strong wood with straight and close grain. Polishes well and withstands wear.	Unsuitable for external use. Narrow planks and tends to warp.	Furniture, floors, tools, turning, toys. Often used for steam bending and laminating.
Iroko	Often used instead of teak because it is cheaper. Very durable. Often used for external work because it is an oily timber and it needs no preservative.	Heavy and cross-grained.	Interior and exterior furniture, flooring and cladding.
African Mahoganies (Sapele, Utile etc)	Inexpensive timber with a plentiful supply. Available in long, wide boards, it is fairly strong and durable.	Properties and colours may vary.	Furniture, shop fitting, floors and cladding.
European Oak	Very strong, durable and tough with little shrinkage.	Very expensive. Japanese Oak is slightly cheaper but less durable. Hard to work with and can corrode iron and steel fittings because it contains tannic acid.	Expensive furniture, veneers, boat building and floors.
African Walnut	An attractive wood that is fairly good to work with. It is available in larger sizes.	Very expensive. Cross grain can make planing and finishing difficult.	Expensive furniture and interior work. Available as veneer.

Hardwood and softwood

- Hardwood comes from trees that are slow growing, which results in a tighter and denser grain.

- Softwood comes from quick growing evergreen trees. It is usually cheaper, softer and easier to work with than hardwood.

The terms softwood and hardwood are used to describe the seed, leaves and tree structure. Therefore hardwood such as balsa is light and soft, whereas softwoods like yew are heavy and hard to work with. On the other hand, hardwood such as sapele are quick growing and cheaper than softwoods such as parana pine.

These woods are available in the following forms:

- Boards: Under 40 mm thick, cut to the required length.

- Planks: Over 40 mm thick, cut to the required length.

- Square and rectangular sections: Common sizes are 25mm × 25mm, 38mm × 38mm and 50mm × 50mm.

Common Softwoods

Name	Advantages	Disadvantages	Uses
Redwood Pine (Fir, Scots pine etc)	One of the most inexpensive and commonly used timbers. It is easy to work, has straight grain and is fairly strong and durable.	It can have a lot of knots and sometime has a blue stain from harmless fungi.	Used for a wide variety of purposes from inexpensive furniture to construction work.
Whitewood (Spruce)	Resistant to splitting, it is easy to work with and is fairly strong.	Contains pockets of resin and small hard knots. Not very durable.	General internal construction work.
Western Red Cedar	A straight, knot-free grain that is lightweight and soft. It is very easy to work with and attractive finish. Contains natural oils that resist insect attack, weather and dry rot.	Not very strong and more expensive than red and whitewood.	Cladding the outside of buildings, paneling walls, kitchens and bathrooms.
Parana Pine	Available in long, wide, knot free boards. Hard, straight grain allows it to be worked easily and achieve a smooth finish. Attractive grain with red and brown streaks.	Can be as expensive as some hardwoods. May warp.	Quality internal joinery work.

Activities

1 What impact does the use of hardwoods have on the environment?

2 Which is the most environmentally sound wood? Explain your answer.

Key points

- All woods are classified into categories.
- The selection of wood is dependent on its properties, aesthetics and cost.

Manufactured boards

Manufactured boards are made by compressing wood particles or thin sheets of wood (veneers) together with adhesive. They have a number of advantages over natural timber:

- They are available in large sheets.
- They are stable therefore they will not warp or shrink like timber.
- They are widely available.
- Decorated veneers or plastic laminate can be easily applied.
- Plywood and blockboard have good strength because of the grain crossing.

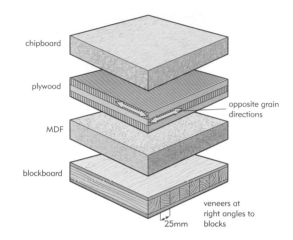

Manufactured boards

Manufactured board

Type	Process of manufacture	Uses	Common Sizes Full sheet 2440 x 1220mm Half sheet 220 x 1220mm
Plywood	An odd number of plies (veneers) are glued together so that the grain of each ply is 90 degrees to the next.	Furniture and interiors when faced with decorative veneers or laminates. External construction work. Ply is graded for interior or exterior use.	4, 6, 9, 12mm thick 3, 5, 7, 9 ply
Blockboard	Blocks of timber glued together and sandwiched between veneer.	The same as plywood but cheaper to produce than ply over 12mm. Usually made for interior use.	18mm thickness
MDF (Medium Density Fibreboard)	Fibres from wood waste are broken down into fine particles. These are mixed with resin adhesive and compressed into sheets.	Furniture and interiors. Excellent surface for veneers and laminates.	4, 6, 9, 12,18mm to 50mm thick
Chipboard	Wooden chips mixed with adhesive and compressed into sheets.	Inexpensive to produce, chipboard is often faced with plastic laminate and used for shelves, worktops and inexpensive furniture.	12 and 18mm thick. Work surfaces 40mm thick

Board	Characteristics	Uses	
Medium density fibreboard	MDF is very widely used in mass-produced, flat-packed furniture. It is a very dense material with an excellent surface finish, which can be veneered or painted. It is very stable and is not affected by changing humidity levels.	• Flat-pack furniture • Drawer bottoms • Kitchen units • Heat and sound insulation	
Plywood	This board is made up from an odd number of layers (veneers) normally 1.5 mm thick. The grain of each layer is at right angles to the layer either side of it. The two outside layers run in the same direction. Plywood is very strong in all directions and it is also very resistant to splitting. Birch veneers are usually used on the outside layers resulting in an attractive surface.	• Boat building (exterior quality plywood) • Drawer bottoms and wardrobe bottoms • Tea chests • Cheaper grades used in construction industry for hoardings and shuttering	
Chipboard	This board is made from waste products. It is bonded together using very strong resins. Although it has no grain pattern, it is equally strong in all directions but not as strong as plywood. The surface is generally veneered or covered with a plastic laminate. It is not very resistant to water although special moisture resistant grades are available.	• Large floor boards and decking for loft spaces • Shelving • Kitchen worktops • Flat-pack furniture	
Blockboard	This board has long strips of timber running down its length. A veneer is applied to the external surface. It does not possess uniform strength.	• Fire doors • Tabletops	

Activities

1 Explain why it is more environmentally friendly to use MDF veneered with hardwood than solid hardwood.

2 Make a list of furniture that is made from manufactured board around your school.

Key points

● Manufactured boards are inexpensive to produce and purchase

● Manufactured boards are more frequently used in the production of furniture than solid timber, because they are inexpensive and easy to work with.

Sawing and sanding wood

Woods can be cut, sawn and sanded using hand tools and machines. Using machines will not necessarily be quicker or easier, therefore you will need to be aware of a variety of methods and when to use them.

Sawing

Hand tools

To obtain a straight cut the following tools can be used:

- Rip saw: This cuts along the grain.

- Cross-cut saw: This cuts across the grain. It can also cut along it but at much slower speed than the rip saw.

- Tenon saw: This produces accurate cuts in smaller pieces of timber. It is used to cut joints.

To obtain a curved cut the following tools can be used:

- Coping saw: This cuts on the back stroke and the blade can be adjusted to any angle. It is usually used on thinner wood or to remove waste from joints.

- Pad saw: This is used to cut curved shapes in the middle of a panel where a coping saw cannot be used.

Machinery and electrical hand tools

To obtain a straight cut the following machines can be used:

- Circular saw: This is a versatile machine that can cut with the grain, across the grain and at any angle to the grain. It is used to cut large boards or panels down into sizes for construction.

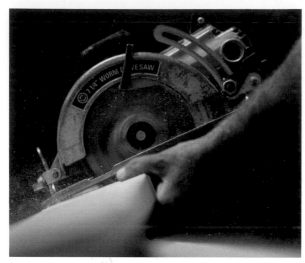

A circular saw

To obtain a curved cut the following machines can be used:

- Band saw: This saw can be fitted with a variety of blades which will cut metal, wood and plastic. Depending on the blade being used it will cut both curved and straight lines. The work is brushed against a continuous metal band with teeth on one edge.

- Jig saw: This is a hand held power tool which is portable and can cut into the centre of large sheets if a hole is drilled to allow the blade to pass through.

- Fret saw: This is a bench mounted saw which can be used to cut very tight curves. It is used to cut thin sheet material.

Sanding

- Belt sander: This can be either a machine or a hand held tool. A continuous abrasive band rotates on two rollers to remove material quickly.

- Disc sander: This can be vertically mounted on a machine or a hand held power tool that can be moved across a larger surface. A disc of abrasive paper rotates to remove material.

- Orbital finishing sander: This is a hand held sander which has an abrasive material stuck or clipped to a pad. Tiny rotations are made which produces a smooth finish. It is used to create a smooth surface before painting or varnishing.

An orbital sander

Router

A router is a handheld device which can also be bench mounted. It is a very versatile machine that can perform a wide range of tasks. It has a rotating cutter, called a router bit, which can cut and shape wood. Router bits come in a variety of shapes and sizes, which can easily be changed for different tasks.

Router bits fall into three categories:

- Cutting to shape – circles, curves and straight lines

- Cutting joints – dovetails, rebates, etc.

- Decorative moulding – round and shaped edges as well as raised and rebated panels.

Health and safety

The correct use of machinery and electrical hand tools is extremely important. They have the capability to seriously injure the person using the device as well as people nearby. It is therefore essential that they are used correctly. Pupils are not allowed to use machines such as circular saws and routers – only people who have been trained to operate them can use them. Other machines may also be prohibited from pupil use. Your teacher will demonstrate how to use the tools and machinery safely. They will also tell you what precautions, procedures and protective equipment should be used. These rules must be followed for the protection of yourself and others.

Activities

1 When should you use a:
   - cross-cut saw?
   - coping saw?

2 What is the difference between:
   - band saw and a jig saw?
   - disc sander and an orbital sander?

Key points

- The selection of the correct saw to use is dependent on the type of cut required, the size of the wood and the type of the wood.

- The selection of the correct sander to use is dependent on the size of the wood, the amount of wood to be removed and the finish desired.

Cutting wood

Planes and chisels are used to cut and shape wood.

Planes

Planes are used to shape wood. They use a fixed blade to remove small amounts of shavings. There are a variety of planes that are used for specific purposes.

Bench planes are commonly used and they come in three sizes:

- Smoothing plane (length = 250 mm): This is a general purpose plane for removing tool marks and making surfaces smooth. It is also used for planing end grain.

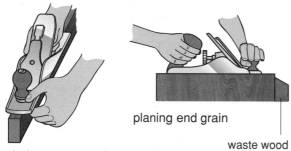

planing end grain

waste wood

planing a narrow edge

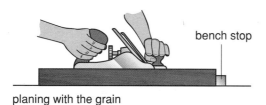

bench stop

planing with the grain

Planing

- Jack plane (length = 350 mm): This is used for planing wood to size. Because of its length it produces a flatter surface.

- Fore plane/jointer plane (length 450/600 mm): This is used to straighten up long edges and flatten surfaces.

Specialist planes are also used and these include the following:

- Plough plane: This is used for cutting grooves and rebates parallel to the edge.

- Rebate plane: This is used for cutting rebates parallel to the edge. A rebate is a section of wood that has to be removed.

- Router plane: This is used to level out bottom joints and rebates that are parallel with the surface.

- Shoulder plane: This is used for cleaning up rebates, halving and shoulders of joints. The width of the blade allows it to get into the corners of joints.

- Bull nose plane: This is the same as the shoulder plane but it has a short nose at the front allowing it to get up to an obstruction.

- Spoke shave plane: This is used to smooth curves. There are two types; one for the inside of curves (concave) and one for the outside of curves (convex)

How to plane

Before planing you need to check that the plane is set correctly by looking along the flat bottom (sole) of the plane.

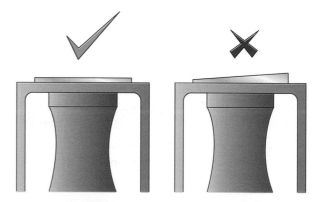

The sole of a plane

- The blade should stick out a distance equal to the thickness of a piece of paper. The blade can be moved in and out using the adjustment screw.

- The blade should stick out equally along the sole. This can be adjusted by the lateral lever, which is above the adjustment screw.

- Test on a piece of scrap wood to ensure a paper thin shaving is produced. Adjustments can be made before planing the final piece.

To plane, press down on the front of the plane when you start to cut, and when it is at the end of the cut press down the heel of the plane.

Chisels

Chisels are used to cut wood when a plane cannot be used. They are primarily used to cut joints. There are three types of common chisel:

- Firmer: A general purpose chisel.

- Bevel edged firmer: This has a bevel on the blade allowing it to be used to cut accurately into corners such as dovetails.

- Sash mortice: This is used for heavy duty work such as mortice joints and levering out waste.

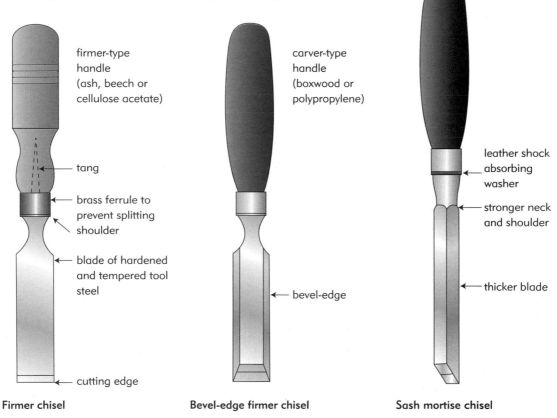

firmer-type handle (ash, beech or cellulose acetate)

tang

brass ferrule to prevent splitting shoulder

blade of hardened and tempered tool steel

cutting edge

Firmer chisel

carver-type handle (boxwood or polypropylene)

bevel-edge

Bevel-edge firmer chisel

leather shock absorbing washer

stronger neck and shoulder

thicker blade

Sash mortise chisel

Types of chisels

Activity

1 Sketch a plane. Use annotation and text to describe how to set it up correctly.

Key point

- Wood can be quickly, easily and accurately removed by either a plane or a chisel.

Joints in wood

There are a wide variety of joints that can be used for connecting wood. You will need to consider a number of points in deciding which to use, including strength required, time and difficulty involved, visual appearance, and tools and equipment needed.

Temporary joints

Knock-down fittings (KD)

These are commonly used in modern furniture. They allow the product to be taken apart for storage and transportation and assembled later. Knock-down fittings provide simple constructions for difficult materials such as chipboard and are advantageous for manufacturers because less storage space is required and this means that time is saved in manufacturing. Knock-down fittings are quick and easy to manufacture, requiring limited machinery and limited skilled labour.

Modesty blocks and bloc-joint fittings

- A modesty block is a plastic block that can be screwed at 90°. It is used for light construction or strengthening joints.

- A bloc-joint fitting is similar to a modesty block but comes in two halves for easy assembly.

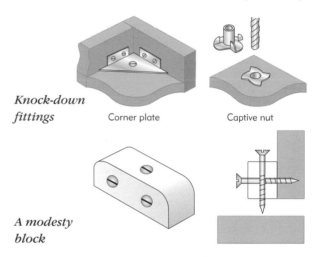

Knock-down fittings   Corner plate   Captive nut

A modesty block

Name	Advantages	Disadvantages	Frame	Box
Butt	Easy and quick to make.	Lacks strength but screws or nails can be added to improve strength. End grain can be seen.	x	x
Mitre	No end grain can be seen – good appearance.	Requires great accuracy to mark and cut. Lacks any real strength but dowels can be used to strengthen it.	x	x
Dowel	Quick to produce, strong.	Needs to be accurately marked and drilled.	x	x
Cross halving	Good strength, neat finish.	Chiseling is required. Creating a good fit can be time consuming.	x	
Corner halving	Quick and easy.	Must be accurately marked and cut.	x	
Housing	Easy to produce, strong.	Achieving a good fit can be time consuming.		
Mortise and tenon	Extremely strong with a neat appearance.	Time consuming. Involves a wide variety of processes. Must be accurately marked and cut.	x	

Scan fittings

● These are used for both frame and box constructions. They consist of a dowel joint or mortice and tenon, which locate the two pieces of material to be fixed. A screw passes through the fitting and locates with a thread insert. When this is screwed in place it tightens the joint.

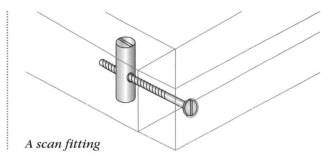

A scan fitting

Adhesives

Material	Wood	Metal	Acrylic	PVC	Expanded polystryrene	Rigid polystryrene	Melamine/ plastic laminate	Ceramic/ glass	Fabric	Leather	Rubber
Wood	PVA										
Metal	Epoxy Contact	Epoxy									
Acrylic	Epoxy Contact	Epoxy Contact	Tensol								
PVC	Contact	Contact	PVC cement	PVC cement							
Expanded polystryrene	PVA	PVA	PVA	PVA	PVA Copydex						
Ridgid polystryrene	Contact	Contact	Contact	Contact	PVA	Polystryene cement					
Melamine/ plastic laminate	Contact	Contact	Contact	Contact	PVA	Contact	Contact				
Ceramic/ glass	See below	Epoxy Contact	Epoxy Contact	Epoxy Contact	PVA	Epoxy Contact	Epoxy Contact	Epoxy			

Adhesives

It is important that the correct adhesive is used for the material that is being fixed.

Do not glue glass to wood. Wood shrinks/expands/warps and this will cause the glass to crack.

Tips

● Surfaces to be joined must be dry, grease free, clean and have no old paint, varnish or glue etc.

● Some adhesives, such as PVA, require pressure to be applied while drying. Before the adhesive is applied, see how the parts fit together and determine what the best method for joining them is.

● Always read the manufacturer's instructions.

Activities

1 What are the advantages of using KD fittings?

2 Why do surfaces have to be prepared before applying adhesive?

3 What joint could be used to make a:
● picture frame?
● desk drawer?

Key points

● Knock-down fittings are widely used and offer advantages over conventional joints.

● The selection of an adhesive is dependent upon the material being fixed.

Preparation of wood

It is essential to clean and prepare wood before a finish is applied. A bad finish can spoil a well constructed product. The preparation process requires a lot of time, care and attention. Finishes not only add to the aesthetic appearance but are also used to protect the product making it durable for use.

Abrasive papers

- A coarse grade glass paper is used first to remove scratches, pencil marks and excess glue. Do not use an eraser to remove pencil marks, this may show through the applied finish along with any glue that has not been removed (to prevent excess glue sticking to the product, masking tape can be used). The glass paper should be wrapped around a cork or wood block and moved along the surface of the wood in the direction of the grain. If MDF is being used try to keep the movement in one direction.

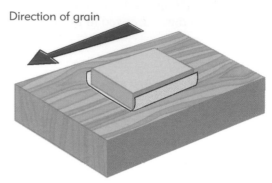

Direction of grain

Glass paper must be used in the direction of the grain

- A medium grade paper is then used in the same way.

- A fine grade paper is then used. The surface should be checked to ensure that all marks have been removed and the product has a consistently smooth finish.

A range of abrasive papers

Abrasive Papers

Grit Size		Glasspaper
220	Extra fine	Flour 00
150	Fine	0 1 1.1/2
80	Medium	F2 M2 S2
40	Coarse	2.1/2

- Any dust left from sanding must be removed. Wipe off as much as possible with a cloth and then use a soft brush, vacuum or tack cloth. A very small amount of methylated spirits can be applied to a cloth to wipe off any fine dust that remains.

- When applying the finish remember that a number of thin coats are required rather than one thick coat. By applying a number of thin layers, a smoother more even finish will be achieved. In between coats the surface can be lightly sanded with a fine glass paper.

Wire wool

Wire wool is available in different grades. They range from coarse (5) to very fine (000) for final finishing. Wire wool is very resilient to clogging and will follow the contours of complex mouldings and shapes.

Mechanical sanding

Mechanical sanders have the advantage of allowing the work to be prepared quickly. They do have drawbacks though. Disc sanders will sand across the grain and leave circular marks.

Portable disc sanders are often used to remove wood, paint and varnish quickly. Belt sanders can be used at the beginning of preparation before medium and fine hand sanding. They can quickly sand down veneers. However, they can leave ridges on the surface, especially in the case of portable machines. Orbital sanders, on the other hand, do give a good finish, but they will not remove scratches and indents quickly. They should be used with medium and fine paper for a good final finish.

Final preparation

- Before applying the finish, prepare a space for yourself and your work. It should be well-lit, ventilated and clean.

- Put down old newspapers to protect the work surfaces.

- Make sure that the applicator is clean (brush, rag, nozzle, etc.)

- Ensure that you have enough of the finish to complete the job.

- Have spare rags, mixing sticks and containers to hand just in case.

Activity

1 Describe the process you would need to carry out to prepare and apply a matt varnish to a shelf made from beech.

Key point

- Time needs to be spent preparing the wood before any finish is applied. The better the preparation the better the finish.

Wood finishes

Finishes are available in water based and spirit based forms. A water based finish can be diluted with water, and brushes can be cleaned with water. Spirit based forms can be diluted with white spirit, and brushes will need to be cleaned with white spirit. Finishes come in matt, satin and gloss to achieve different effects. Most finishes are available for interior and exterior use.

Paint and varnish

Paint and varnish can be applied using a brush or they can be sprayed on. Most forms of spirit-based paints and varnishes can be obtained in a spray form, but an air compressor and spray gun can be used to apply any type of finish. Spirit based polyurethane paint and varnish are often used when a tough, durable surface is needed. Before applying the paint, a primer and an undercoat are needed as they help to seal the wood before the top coat is applied.

Polish

Polish is usually used on furniture to give an attractive finish. It is not as hard wearing as varnish but is a traditional way of finishing wood. The two most common types of polish are beeswax and shellac.

Shellac is made from crushed beetles and is dissolved in methylated spirit. The first coat is applied with a brush to seal the grain, then it is lightly rubbed down along the grain with a soft cloth. After the first coat is dry, more shellac can be applied and rubbed in small circles. This is called French polishing.

To apply beeswax the wood must have a sealer applied using a brush. After it is dry it will need to be rubbed down with wire wool. The beeswax can then be applied using a soft cloth to rub it in. Beeswax can also be used on top of varnish. After the final coat has dried the beeswax can be rubbed in using wire wool.

Beeswax and shellac

A wide range of wood stains is available

Stain

Stains apply colour to the wood but allow the grain to be seen through. Staining is done before a final finish is applied. Stains are now available that stain and finish all in one and come in a wide range of colours.

Preservative

Exterior wood will need extra protection from decay. Preservatives soak into the wood and protect it from fungus and insects as well as the weather. The most commonly used preservative is creosote; this leaves a matt brown finish and is inexpensive to buy.

Activity

1 Give an example of a product which has each type of finish and comment on why this finish would be best suited for it.

Key point

● Finishes are used to add to the aesthetic appeal of a product as well as fulfil a preservative function.

Plastics

Plastics are produced from oil-based petrochemicals. This raw material is refined in a chemical plant to produce plastic. This base plastic is then processed to produce the final product.

The use of plastic is a major environmental issue as its origin (oil) is a non-renewable source. The disposal of plastic is also a concern as it is not biodegradable and it is difficult to recycle due to the wide variety of plastics used in products.

This problem has been addressed by the invention of biodegradable plastic made from organic matter. Many products are now labelled to identify what type of plastic has been used so it can be recycled easily. There are two main groups of plastics as outlined below.

Thermoplastics

Thermoplastics are made up from long chains of molecules, called polymers, with very few cross linkages, called monomers. When heated the molecules move and become detangled.

Common Thermoplastics

Name	Form	Properties	Uses
Acrylic (PMMA)	Rods, tubes, sheets in a range of colours and glass clear.	Easily machined and polished. Very durable with high impact resistant but scratches easily.	Signs, furniture, light units, watch and clock glass, skylights, and baths.
Polypropylene	Powder, granules, sheets, rods in a range of colours.	Very light with good resistance to wear (e.g. hinges). Good mechanical and electrical properties.	Containers, folders, lampshades, medical equipment, rope, sacks.
Nylon	Powder, granules, chips, rods, sheet, tubes, extruded sections.	Tough and hardwearing with resistance to friction.	Gear wheels, bearings, packing, clothing, automotive equipment.
Polystyrene	Powder, granules, sheets in a range of colours.	Stiff and hard but it can also be brittle.	Toys, containers kitchenware, disposable plates and cutlery.
Expanded polystyrene	Beads, sheets, slabs.	Lightweight and buoyant. Absorbs shocks but crumbles and breaks easily.	Packaging sound insulation and heat insulation.
Low Density Polythene	Powders, granules, film, sheets in a range of colours.	Tough, flexible and soft. A good electrical insulator.	Bags, sacks, bottles, cable insulation.
High Density Polythene	Powders, granules, film, sheets in a range of colours.	Stiff and hard with good chemical resistance.	Buckets, bowels, bottles and pipes.
Rigid PVC (polyvinyl chloride)	Powders, pastes, liquids, sheets.	Lightweight, tough and stiff.	Pipes, containers, roofing sheets.
Plasticised PVC	Powders, pastes, liquids, sheets.	Soft and flexible with good electrical insulation.	Vinyl tiles and wallpaper, hosepipes, dip coating.

This makes the plastic flexible and easy to mould and form. When the plastic cools the molecule chains reposition themselves and become tangled again. This results in the plastic returning to a stiff state once again. Thermoplastics therefore have what is known as a 'plastic memory'. This means that thermoplastic can return to its former shape after being heated and cooled.

Thermoset plastics

Thermosets set with heat and therefore have little plasticity. The molecules in thermoset plastics link both side to side and end to end, which makes for a very rigid material that cannot be reheated and changed. This makes them ideal materials for manufacturing products that need to maintain their shape and resist heat.

Common Thermosetting Plastics

Name	Form	Properties	Uses
Urea formaldehyde (UF)	Powder/Granules	Stiff, hard and heat resistant but also brittle.	Plugs, sockets, switches.
Phenol formaldehyde (PF)	Powder/Granules	Stiff, hard and heat resistant but also brittle.	Saucepan handles, door handles, electrical fittings.
Melamine formaldehyde (MF)	Powder/Granules	Stiff, hard and heat resistant as well as stain resistant.	Electrical installation, furniture – kitchen worktops, tabletops.

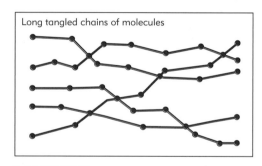

Long tangled chains of molecules

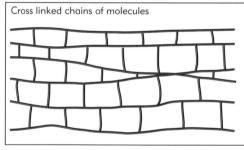

Cross linked chains of molecules

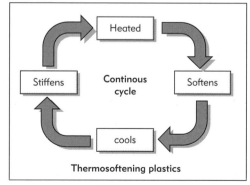

Thermosoftening plastics

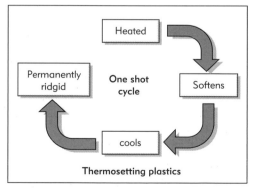

Thermosetting plastics

The structure of thermosoftening and thermosetting plastics

Activity

1 What are the advantages and disadvantages of using:

- thermoset plastics?
- thermoplastics?

Key point

- Plastics can be classified into distinct categories that determine how they are used and processed.

Deforming plastics

Thermoplastics, due to their natural properties, can be easily shaped or deformed. They can be heated in sections or as a whole. There are three basic methods that can be used. For each method a jig or mould is made for the plastic to form around.

Line bending

This process heats a strip across the plastic that will allow for a single bend to the angle required.

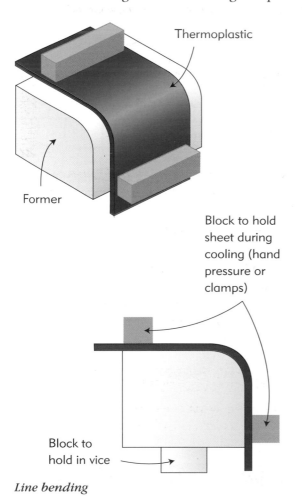

Line bending

- Mark a line across the plastic where it is to be heated.

- Place it over the strip heater so the line is directly over the slot where the heat is coming from. The supports the plastic is placed on can be adjusted to suit the thickness of the material.

- Turn the plastic at regular intervals to ensure it is heated evenly and does not bubble. When the plastic becomes flexible it can be placed on a former.

- The plastic needs to be held or secured in place. Once it is cool it will become rigid and can then be removed.

Press moulding

This process involves the whole of the plastic sheet being heated until it is flexible. This can be done in an oven or under infrared heaters. The temperature should be around 150–180°C for acrylic sheet, as temperatures over 180°C will damage it.

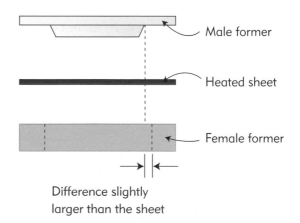

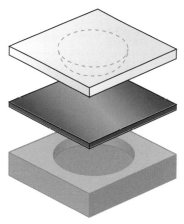

Press moulding

Once it has become flexible with a rubbery quality, it can be placed over the female former. The male mould is then placed on the former and pressure is applied to bring the two sides of the mould together. The pressure needs to be maintained until the plastic has cooled.

Vacuum forming

- The mould is placed on the platen and the thermoplastic is clamped above it.

- The plastic is heated until it softens and slightly sags in the middle.

- The platen is raised and the vacuum button is pressed to suck out all of the air. The plastic should form around the shape.

- Once the plastic is cooled the mould can then be removed.

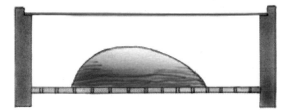

The mould is placed in the vacuum forming machine and a thin sheet of plastic is secured above it.

The plastic is heated until it starts to sag.

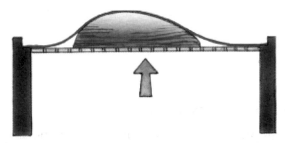

The vacuum pump is turned on. This sucks out all the air between the plastic and the mould.

The plastic is sucked down over the mould. When the plastic is cool it sets into the shape of the mould.

Vacuum forming process

Instead of the air being sucked out it can also be blown in forcing the plastic into the mould. This is often used in industry to produce products like bowls and baths. This process is called blow moulding.

Activity

1 Identify a product that has been made from each of the deforming methods.

Key point

- Deformation processes use moulds to shape heated thermoplastics.

Reforming plastics

Plastics come in a base form – powder, resin, granules, etc. From this they can be processed and shaped in a mould to form a desired shape. When the plastic has set or cooled it maintains the form of the mould. Therefore the plastic has been reformed to make the product.

Injection moulding

Injection moulding is one of the most common methods used in industry for producing large quantities of the same item. It has enabled products to be produced cheaply and with precision.

The initial tooling costs are high. Making a mould can be expensive but it can produce thousands of the same items, which makes the cost of each unit very low. Moulds are made in two or more parts that can be separated to allow the product to be easily removed. For high production runs moulds are made from steel. They can also be made from aluminium for shorter production runs.

The process involves granules of thermoplastic being fed into a heated machine, which reverts the plastic to a liquidized state. It is then quickly forced under high pressure into the heated mould. The mould is heated to allow the plastic to remain in a liquidized state so it can run throughout the mould and does not harden before the mould is full. Once cool the mould can be separated to release the product.

Extrusion

Extrusion works on a similar method to injection moulding. It allows a continuous shaped section to be produced such as pipes and UPVC window frames. The plastic granules are fed into a heated machine and onto a rotating screw. The heat reverts the granules to a liquidized state and the screw forces the plastic under pressure through a die (the shape of the product). Once it is passed through the die it is quickly cooled to retain its shape.

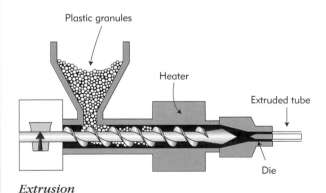

Extrusion

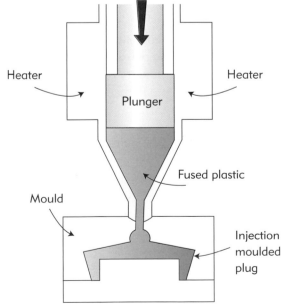

Injection moulding of a plug

GRP (glass reinforced plastic)

GRP is often referred to as fibreglass. It is an extremely strong and lightweight material that does not corrode. It is used to produce products in a variety of shapes and sizes from car bodies to seats to boats.

The process involves woven fibre matting being laid onto a male or female mould and covered in polyester resin. This process is continued until the desired thickness is achieved. Areas that require extra reinforcement can be built up and strengthened. When the polyester resin has set, the mould can then be removed.

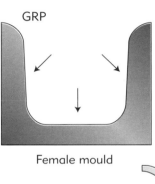

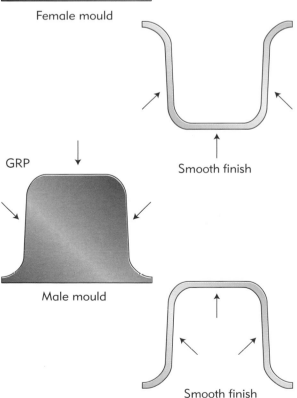

The process of forming GRP

The woven fibre matting used is usually glass fibre. Other types of matting and resin are also used that offer properties that glass fibre and polyester resin do not. An example of this would be the use of modern materials such as Kevlar, which offers high impact resistance for products such as helmets.

The production of the mould is also important. As with vacuum forming the mould has to be shaped to enable easy removal. The surface of the mould has to be smooth to give the product a smooth finish. Only the side of the GRP product that comes into contact with the surface of the mould achieves a good finish.

Activities

1 Why is injection moulding a commonly used process in industry?

2 When would you use extrusion instead of injection moulding?

3 What are the advantages of GRP moulding?

Key point

● Reforming methods allow a wide variety of shapes to be produced repeatedly.

Mechanical systems

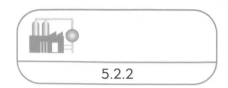

In mechanical systems four different types of motion are used:

- Linear: Where the object moves in a straight line.
- Reciprocating: Where the object moves in a straight line backwards and forwards.
- Rotary: Where the object moves in a circular motion around a pivot.
- Oscillating: Where the object moves in a circular motion clockwise and anti-clockwise around a pivot.

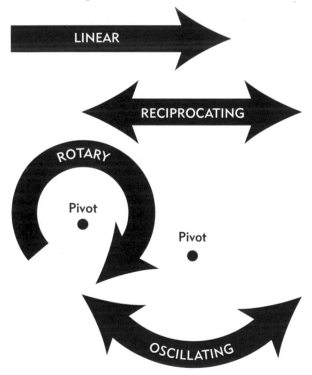

The four types of motion used in mechanical systems

Mechanisms are used to change the input motion to a desired output motion. Different types of mechanisms are used to achieve a variety of outputs. Mechanisms can change motion by:

- speed

- force
- direction.

For example, in a hand whisk the input motion is changed as shown in the table below:

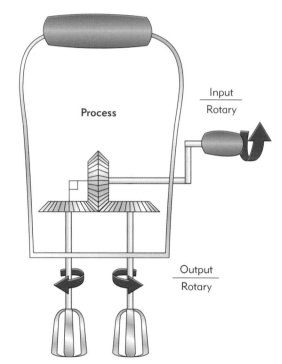

A hand whisk

Input	Process	Output
Rotary motion from the handle	A bevel gear turns the motion through 90° and changes the speed	The whisk rotates faster than the input

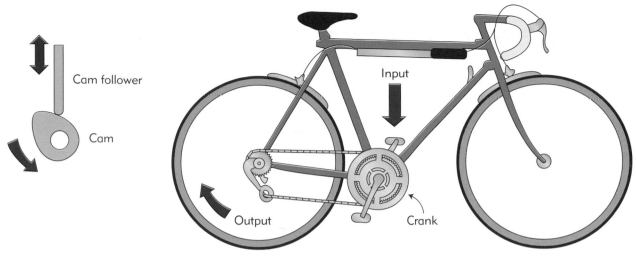

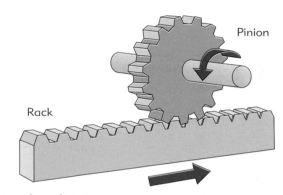

Cams and cranks

Cams and cranks

- Cams change rotary motion into reciprocating motion. They are fixed off centre to a rotating axle with the follower resting on the edge; this follows the profile of the cam. Different shaped cam profiles can offer a variety of movements and these are often used in toys.

- Cranks work on a similar system to that of a cam. They can convert rotary motion into reciprocating motion, but can also convert reciprocating motion into rotary motion. The most familiar crank is used on a bicycle. The linear force of the legs (input), converts to a rotary motion of the wheel (output).

When cranks are joined together they form a crankshaft. A crankshaft is used in a car engine. The reciprocating motion of the piston rotates the crankshaft and this in turn rotates the wheels. This is known as a crank and slider as the piston slides up and down its casing.

Rack and pinion

To change rotary motion to linear motion a rack and pinion are often used. The pinion is fixed to a rotating shaft and the rack is fixed to a surface. Both the rack and pinion have teeth which mesh together so when the one moves the other moves with it.

A rack and pinion

Activity

1 What is the input, the process and the output for the following:

- a door handle?

- a bicycle?

Key points

- Mechanisms can change the input motion into a different output motion.

- The correct mechanism must be used as a process to produce the desired output motion.

Gears

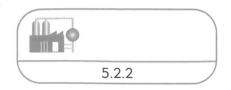
A gear is a mechanism that is often placed on a rotating shaft. It can be used to change motion by speed, force and direction. Gears are a widely used mechanism in vehicles and machinery.

Spur gears

A spur gear is a disc with teeth around the rim. The teeth allow the gears to be meshed together so when one rotates the other will rotate with it. They also prevents the gears from slipping.

The gear receiving the power from the motor or hand (input) is called the driver gear. As it rotates, the gear meshed with it will rotate in the opposite direction; this is called the driven gear (output). If the driver and the driven gears need to rotate in the same direction a smaller gear is placed in between them; this is called an idler gear. The idler gear does not affect the speed or force but is placed there to change the direction of rotation.

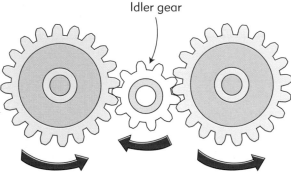

An idler gear

As well as changing direction, spur gears can also change speed and force. We can work out the number of times the driven gear (output) will turn using the following formula:

Number of teeth on the driver gear (input)

―――――――――――――――――――――――――

Number of teeth on the driven gear (output)

For example:

- If the driver gear has 10 teeth and the driven gear has 5 teeth:

    10 [divided by] 5 = 2.

- This means that when the driver gear turns one full rotation the driven gear will rotate twice.

- This is often written as 1:2 and is called the Velocity Ratio (VR).

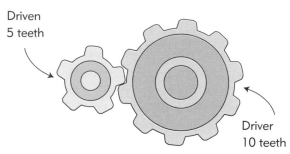

The velocity ratio is 1:2

Gear trains

Gear trains are often used to transform a high speed input (low torque) into a low speed output (high torque) to create more force. They can also be used the other way round to create a high speed output.

Pairs of gears are fixed to the same rotating shaft. These are then meshed with similar pairs of gears. This can be continued to create the desired speed and force.

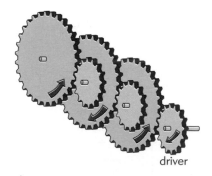

A gear train

For example:

- If the driver gear turns the first larger gear, the smaller gear connected on the same shaft will turn at the same speed. This gear is meshed with the next larger gear and so on.

- To work out the output speed all the individual VRs are worked out and multiplied.

- The VR for the driver to the first driven gear is 2:1 (the driver gear has to rotate twice for the driven to rotate once).

- The next is 3:1 and the last one is 4:1. Therefore the final output is worked out by:

$$2:1 \times 3:1 \times 4:1 = 24:1$$

- This means then that the input gear has to turn 24 times in order to turn the output gear once.

It is important to remember that if the output speed is slower than the input speed it will turn with more force. Conversely, if the output speed is faster than the input speed it will turn with less force.

90° Transmission gears

Spur gears and gear trains do not allow the plane of motion to be turned through 90°. In order to do this other types of gears have to be used. The following gears are commonly used to do this.

Bevel gears

Bevel gears slope at a 45° angle, so when two are used together they make up 90°. The teeth of the gears are on the sloping face so they must be used with this face sloping inwards. If they are not then the gears will not work.

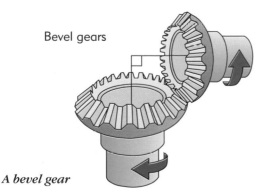

Bevel gears

A bevel gear

Worm gears

Worm gears can only be used as an input (driver) gear and are used with a spur gear for the output. They are often used to drastically reduce speed because a worm gear has only one tooth. When the worm gear rotates once it turns the gear meshed with it by one tooth only. For example, if the spur gear meshed with the worm gear has 20 teeth, the worm gear must rotate 20 times in order for the spur gear to rotate once.

A worm gear

Activities

1 a If the input gear has 42 teeth and the output has 6 teeth, what is the VR?

   b How many times would the output rotate if the input rotated:

   - 10 times?
   - 35 times?
   - 50 times?

Key points

- Spur gears are used to change the direction, speed and force of the input motion.
- The output can be predicted by working out the VR.
- Motion can be transferred through 90° using bevel and worm gears.

The control device

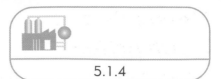

As part of your GCSE coursework, you are required to produce a control device to manufacture your product. A control device can be a template, mould, pattern, jig or former. The type of control device you will use is obviously dependent upon the design of your product. It is important to remember that the control device should offer repeatability, accuracy and ease of use.

Repeatability

The control device must be designed and made so that it can withstand repeated use. This allows the product or parts of the product to be manufactured in quantity. Developing a product that allows repeated production simulates production methods used in industry. It demonstrates the fact that your design is not a one-off, but has been designed for multiple productions.

Accuracy

The control device must maintain its accuracy despite being used repeatedly. The accuracy of the control device should not deteriorate after it has been used once, as this would defeat the purpose of producing a control device.

Ease of use

The purpose of the control device is to accurately reproduce the product, or parts of the product, quickly and easily. The work you have invested in the design and manufacture of the control device should ultimately save you production time. If it is quicker to make the product, with the same degree of accuracy, from scratch, then there is no point in producing a control device.

Successful production

The success of the control device is dependent upon the design, materials and construction. Ask yourself the following questions:

- Will the design allow me to produce the product or parts easily, quickly, accurately and repeatedly?
- Will the materials allow me to repeatedly manufacture without deterioration in quality or loss of accuracy?
- Will the construction withstand repeated use?

Carrying out testing is important to achieve a successful control device. Through testing the device in the way it has been designed to function, you will be able to evaluate its strengths and weaknesses. Modifications can then be made so that it functions correctly.

Materials and construction

The control device is a purely functional device, created solely for the purpose of production. The materials and construction methods used will obviously reflect this and therefore be fit for the intended purpose. The device does not have to look aesthetically pleasing; in fact it may look very crude. Try to use:

- inexpensive materials (mild steel, MDF, chipboard, etc) or off-cuts. This is especially true when producing large control devices

- materials which are easy to work with and shape, such as jelutong, for patterns and moulds

- quick and easy construction methods such as screw, bolts, brackets. These types of semi-permanent fixings will also allow for quick and easy modifications.

For example, I need to produce a right-angled jig to hold acrylic that has been heated on a strip heater. It would be a time consuming and expensive to use finger joints and oak. The same right-angled jig could be achieved by using a pine butt joint strengthened with screws. It would produce the same result but it would be quicker and cheaper.

Finishes

Some patterns, formers and moulds for such processes as GRP, injection moulding, aluminium casting, etc., the surface finish is very important. Lumps, marks and indents on the control device may show up on the product being produced. To maintain the quality of finish and ensure that this does not happen, a smooth finish must be produced. This will also save time on finishing the product later.

For wooden based control devices, work through the grades of glass paper from rough to smooth. Car body fillers can be used to smooth out indents as well as filler spray paints. Sealers can be applied to achieve a smoother finish but remember that paint and varnish may come away from the wood and affect the finish of the product.

For metal based control devices, rough to smooth grades of files and silicone carbide paper should be used. No finishes need to be applied but remember to keep the metal free from corrosion.

Activities

1 Explain why the control device should allow for:
   - repeatability;
   - accuracy; and
   - ease of use.

2 Why should inexpensive materials be used to produce the control device? Explain your answer.

3 Why must some control devices have a smooth surface finish?

Key points

- A control device is produced to offer repeated production that maintains accuracy and is quick and easy to use.

- The success of the control device is reliant on the design, materials and construction used.

- Testing and modification should take place to improve the device.

- Materials and construction should be fit for their purpose.

Questions

1 Name the most suitable metal to use to make the following products:

   a a hammer

   b a racing bike frame

   c a computer cable

   d a router bit.

2 Explain the difference between an alloy and a pure metal.

3 What forms of metal can be cut using a hacksaw?

4 Explain the difference between a milling machine and a lathe.

5 List three advantages a CNC milling machine has over a hand operated milling machine.

6 For the following products, name the wood or metal that would be most appropriate to use:

   a a toy truck

   b a veneered table top

   c a garden bench

   d an inexpensive coffee table.

7 What tools and machinery can be used to cut curved lines?

8 When would you use the following tools:

   a jack plane?

   b spoke shave plane?

   c sash mortice chisel?

9 Explain what a knock-down fitting is and when it would be used.

10 Name the most appropriate plastic to use for each of the following products:

   a a washing up bowl

   b a door handle

   c a light switch

   d a shop sign.

11 What types of processes would be used to produce the following products:

   a a bath plug?

   b pipes for guttering?

   c a washing-up bowl?

   d a canoe?

12 Why is the use of plastics a major environmental concern?

13 Give an example of a mechanism which changes the plane of rotation through 90°.

14 List the four types of motion.

15 What does VR stand for?

PLANNING AND PRODUCTION

Planning for production 1

In order to make a successful product in the time available you will need to have produced:

- a working drawing
- a plan of how you will make it.

The production plan

To ensure a high quality product, detailed planning is needed. Planning is important because it helps you prevent making mistakes and makes good use of your time so that the final product is well made. Through planning you can organize yourself so you know exactly what to do every step of the way. A production plan should break down into four separate plans:

Time plan

A time plan will give you an overall picture of what stages need to be completed by the required time. When you have completed each stage you can colour it in. This will give you a complete picture of what has been done and what needs to be done. This can be presented in a Gantt chart, which can be produced in a spreadsheet package.

Flow chart

A flow chart is an easy way to plan the sequence of production. It is a clear way in which you can order how the manufacture will take place. It also allows you to control the process through adjustment.

Activity	Lesson									
Time plan for the construction of a jewellery box **All lessons 2 hours** **Total time 20 hours**	1	2	3	4	5	6	7	8	9	10
Mark out sides, lap joints and rebates Cut out and check for fit	X	X								
Glue sides together Measure and mark lid/bottom Cut out and check for fit			X							
Set up CAD/CAM for lid Cut out detailing on lid Glue lid and bottom into rebates				X						
Clean up/sand box Mark where lid is to be cut Cut lid					X					
Mark out rebates for hinges Chisel out rebates					X	X				
Test fit hinges Mark out lock for box Cut/drill to fit lock							X			
Test fit lock Clean/sand inside of box							X			
Cut out velvet for the inside Test fit velvet								X		
Final clean up/sand inside and outside box Vanish box								X		
Sand box down lightly Apply second coat of varnish									X	
Glue in velvet Fit hinges and lock										X

Time plan presented as a Gantt chart

Production schedule

A production schedule is a stage-by-stage plan of the activities you need to carry out. It is more detailed than a flow chart allowing for quality control and quality assurance procedures to be put in place. This will make good use of your time and ensure a high quality product is produced.

Process plans

At certain stages in the production schedule you may have a task that has a lot of detail such as vacuum forming. This process can be presented in detail using sketches and writing on a separate sheet of paper. It will enable you to plan the entire process in detail so you will know exactly what you are going to do. You should write in

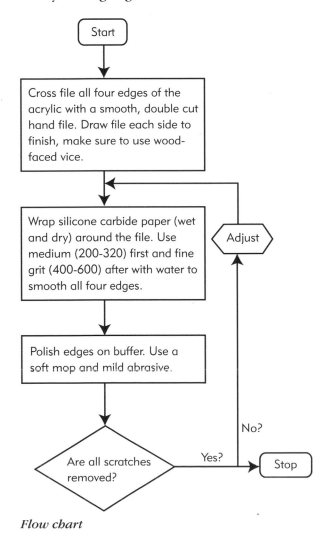

Flow chart

'See process plan' on the production schedule to remind yourself. Process plans can also be shown as flow charts (see Planning for production 2). This allows for feedback to be built in as a quality check.

Pre-planning

Before planning the production some pre-planning will be needed. This may need to be carried out during the development stage of your design, as you may want to alter the design in order for the product to be produced. It will also allow you time to access any materials or equipment you need to produce the product. The following questions are a good guide:

- How many hours or lessons have I got to complete the product?
- Are the materials I want to use available?
- Are the components (fixtures and fittings) I want to use available?
- Can I use pre-manufactured standard components?
- Are the finishes I want to use available?
- Have I got access to the correct tools, equipment and facilities in order to make the product?
- Have I got the skills and the ability to finish the product on time?
- What are the health and safety issues?

Activities

1 What two things do you need to have before you start producing your product and why are they necessary?
2 Why is pre-planning needed?

Key points

- Detailed planning needs to be broken down into four sections.
- Pre-planning should take place in the development stages before the final planning.

Planning for production 2

When planning to make a product it is important to plan the order of manufacture. This should also include the processes, tools and equipment needed in order to complete the tasks. As you will only have a set period of time to complete the production of the product, each stage should have an estimated time for completion. This will help you keep on target.

Flow charts give a quick visual indication of the plan. They are easy to follow with arrows to show the flow of work in the correct sequence.

Flow chart symbols

In order to make flow charts consistent and assist in the visual communication, basic symbols are used. These are called flow chart symbols.

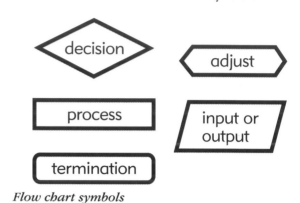

Flow chart symbols

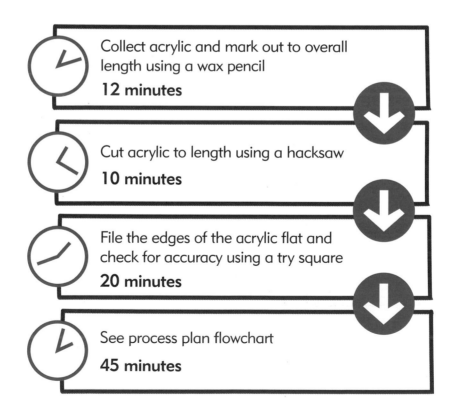

Collect acrylic and mark out to overall length using a wax pencil
12 minutes

Cut acrylic to length using a hacksaw
10 minutes

File the edges of the acrylic flat and check for accuracy using a try square
20 minutes

See process plan flowchart
45 minutes

In using flow chart symbols, feedback and control operations can be included in the planned sequence. This allows you to check and adjust (control) the process so that it is correct before continuing. Whenever a decision needs to be made that will result in a yes or no answer, a loop is put into the system. If the answer is no, the operation is carried out again with adjustments added (the feedback).

This continues until the answer is yes and then the flow chart carries on. This is often referred to as a closed loop system or a system that has built in feedback.

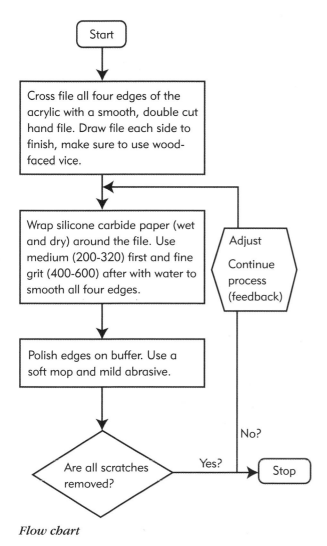

Flow chart

The advantage of flow charts is that they are a useful way of visually communicating the plan of work and sequence of operations. Their disadvantage is that they lack the detailed information needed. Also, because manufacturing products is often a complex process, flow charts can become very long-winded and complex. This is why they are often used in conjunction with production schedules or as process plans.

It is often quicker to draw a flow chart using ICT software.

Activities

1 List two advantages of a flow chart when used to illustrate a process.
2 Draw a flow chart to make a cup of tea.
3 What is:
   • a closed loop?
   • feedback?
4 Why are flow chart symbols used?

Key points

● Flow charts are a good way of visually showing a sequence of events for production but they can lack the detail required for production.

● Feedback in a closed loop can be used to ensure that an activity is correctly done and quality is maintained.

Planning for production 3

Production schedules

Production schedules contain much more information than flow charts. They have quality control and quality assurance procedures built in to ensure that the final product is one of high quality. They also allow you to build in modifications and testing so that evaluations can be made throughout the production process.

Production schedules are often shown as a table or grid. Blanks can be produced on DTP software so that they can be filled in. This saves time and also enables all the required information to be added.

Producing a schedule

You will have to look at the working drawing of your product and break it down into basic stages. Estimate, using your experience, how long each stage will take. Try to give yourself a bit more time than is needed just in case you cannot get access to a machine or you have to modify your product. You may have to adjust the time and activity in order to fit it in to the total time available. This can be done in conjunction with your time plan.

It is helpful to identify where critical stages are in the production schedule. This will serve as a reminder to you as well as allowing for procedures to be included so that they can be achieved. You can do this by writing them in red. The production schedule should include the following for each stage:

Activity – write down what you intend to do, i.e. a brief description of the activity. Make sure you include the following:

- **Tools, machines and processes undertaken**: this should include the resources needed for each stage.

- **Accuracy/size**: this states how close to the accurate size you are prepared to accept. Crucial sizes can be included as a reminder.

- **Time**: how long it will take to complete the task.

Stage or task	Tools, materials, processes to be undertaken	Accuracy required	Time needed, deadline	Quality assurance procedures	Alternative methods of manufacture	Safety – risk assessment

Process Chart 1

	Task	Tools required	Time spent
1	Obtain wood and cut to correct thickness. Measure and mark out wood to correct sizes.	Metal rule, planer, pencil and try-square	1 Lesson
2	Mark out all of the edges that need to be joined together. Using the mitre saw, cut them all at a 45° angle.	Pencil, mitre saw, large clamp (to clamp mitre saw to bench top), small clamp (to clamp wood to mitre saw).	1 Lesson
3	Measure and mark out the wood needed for the lid. Cut all of the required edges to make mitre joints.	Metal rule, pencil, try square, mitre saw	1 Lesson
4	Create templates for stained glass windows.	Computer, access to the internet (to obtain the Celtic structure to make my stained glass window), and word Excel computer program, to enlarge my Celtic design. Also a printer (to print my templates from the computer).	I will do this at home. I estimate that it will take me around one hour to do.

	Task	Tools required	Time spent
5	Paint on the gold outline of the Celtic patterns on both pieces of glass (one for the side of my box and one for the lid).	Gold glass outliner	1 Lesson
6	Paint the stained glass windows	Red and green glass paint, and a thin paintbrush (a soft one to reduce the appearance of brushstrokes).	1 Lesson
7	Cut out a hole for the stained glass window to be placed and cut a small ledge either side where the edges of the glass will be attached to.	Rule, pencil, coping saw, drill with attachment, chisel and sandpaper.	2 Lessons
8	Sand box and lid until extremely smooth	File and sanding machine	1 Lesson

Quality assurance – a statement of how you are going to ensure mistakes will be avoided.

Quality control – how you are going to check for accuracy and quality and how you will test the product.

Alternatives – write down any alternative ways in which you can do the activity or what can be done if it does not go as planned.

Health and safety – note any hazards involved.

Modification/evaluation – this is to be filled in as you carry out the stage. It will allow you to note any changes that might affect the following stages as well as results from tests and comments about how well the stage was done. This can also be used in your final evaluation.

This may seem to be a long-winded approach but the more detail you add the better. It will demonstrate your ability to the examiners as well as assisting in the production of your product. Try to set realistic deadlines and activities which are achievable. A schedule like this helps you to plan your time and use it efficiently.

Activities

1 Using ICT software produce a grid that can be used as a pre-printed production schedule.

2 Why are quality assurance and quality control included in the grid?

3 Why is it important to include modifications in the grid?

Key point

● Detailed production schedules can be produced with the aid of pre-printed grids. These grids identify all the key requirements to ensure a quality product is produced.

Health and safety 1

The designer

When designing and making your chosen product, health and safety is an extremely important issue. As the designer you have a responsibility to the user. It is your responsibility to make sure that the product is safe for the user to constantly use and maintain over its lifespan. Analysing the user's requirements at the research stage will assist in this. A badly designed product can cause injury to the user.

An example of this is toy design for young children. Considerations in the design must be made for how the child will interact with the product. Considerations must be made for things like small parts being swallowed, toxic paints, gaps for fingers to be trapped etc.

When designing the product the materials to be used, the processes involved and any standards will have to be considered.

Materials

The material to be used is often chosen because of its properties and aesthetics. The correct material and finish must be appropriate for the product and what it is to be used for. Other considerations will also have to be made concerning safety issues. The material and the design must ensure:

- no sharp edges exposed or splinters occur
- if it does break that it does not shatter, leave sharp edges or expose dangerous parts
- it protects the user from moving or electrical parts
- the material will wear well over a period of time

- fixtures and fittings are well secured and will not pose any dangers
- the material, fixtures and fittings will not corrode easily
- it is protected against flammability
- the correct finish is fit for its purpose, e.g. non-toxic for children, exterior finishes if the product is to be used outside.

These are the points you need to be thinking about as the designer. They will help you to decide which materials to use for your chosen product.

Activities

1 Why are kite marks and CE marks placed on products?

2 Why is selecting the correct material a health and safety issue?

3 Examine an existing product. List what the designer has done to ensure it is a safe product. Could it be improved in any way?

Key points

- Standards are produced as a guide for the designer to ensure that the product meets the required level of safety for the user.
- When selecting materials, consideration for health and safety is a determining factor.

Children's toys must be safe to play with

Health and safety 2

The manufacturer

Badly manufactured products, which cause injury, will result in the manufacturer being held responsible. This may result in legal action and the manufacturer may well be prosecuted. It will also affect the reputation of the company and have an impact on sales. This could result in the company going out of business. Therefore the manufacturer must test and modify the product at the design and prototype stage, using standards that are set out by organizations such as the British Standards Institute.

Manufacturers also have a responsibility to their workers. The environment that the product is manufactured in will have to meet health and safety requirements by law. The manufacturer is therefore legally responsible for the health and safety of their employees. You school workshop has to conform to these standards as well for the protection of the pupils and the teachers.

The working environment

The legal health and safety requirements in the workplace include responsibility for:

- the correct storage of tools, equipment and materials, including adhesives, chemicals and finishes

- ventilation for extracting dust and fumes

- adequate space around machinery and purpose built areas for such activities as spray painting

- correctly functioning machinery with guards for moving parts and emergency stop buttons

- procedures for accidents, trained first aiders, first aid equipment, fire extinguishers etc.

- facilities for the disposal of waste.

Personal safety

Protective clothing can prevent personal injury

In addition, the health and safety requirements state that:

- training for the correct use of tools, machinery and equipment must be given

- protective clothing such as gloves, goggles, facemasks, earmuffs etc. must be provided and worn where necessary

- barrier creams to protect the skin from becoming irritated when using certain materials and liquids must be provided and used where necessary.

Manufacturers will also carry out risk assessments for each activity and process. By analysing a task, e.g. using a pillar drill, they can implement procedures to minimize the risk of an accident, e.g. chuck guards, goggles etc.

Your school will carry out a similar risk assessment for activities that you undertake. This will set out the procedures for each activity to minimize risk. They will also use the COSHH guidelines for the safe use of adhesives and finishes which give off fumes and can be a risk to the users.

The environment as a whole

Recycle logo

Manufacturers have a responsibility to the wider environment as a whole. Many, but sadly not all, manufacturers assess what impact their working practices have on the environment and attempt to rectify them. There are three basic ways they do this:

Disposal

- The correct disposal of waste materials, e.g. chemicals, oils and by-products of manufacture.

- The use of recyclable materials.

- Instructions and facilities for the product to be disposed of at the end of its life.

Reduction

- Using less harmful or non-harmful materials in the product or the production process, e.g. no CFC's in packaging, fridges and aerosols.

- Reduction or filtration of pollutants into the atmosphere or the surrounding environment.

- Reduction in the amount of material and packaging used in the product through the design.

Recycle

- The use of recycled material in the processes and the product.

- Re-cycling the waste material produced.

Activities

1 Carry out a risk assessment on the use of a pillar drill. List the possible hazards involved in the use of it and then list what could be implemented to minimize these risks.

2 What procedures could be put into place in your school workshop to minimize the effect on the environment?

Key points

- Manufacturers by law have to adhere to health and safety requirements to prevent and minimize risk to their employees.

- Manufacturers affect the environment as a whole and procedures can be implemented in order to minimize their impact.

Industrial applications 1

Production methods

Products are manufactured using specific production methods. Decisions have to be made about how a product will be manufactured and how many will need to be produced. This is usually decided at an early stage and included in the specification. The product being produced will often determine which method is used and will have an implication on the cost. The type of production can be one of three categories:

1 One-off production

2 Batch production

3 Mass production.

One-off production

This is a very labour intensive and costly method. One or a small group of people produce, from beginning to end, a highly crafted product.

These products are produced one at a time and are usually original pieces for a specific function. This is sometimes referred to as bespoke production where the product is made to a specific requirement for the user.

Batch production

A team of people produces a required number of the same item. This is done in less time than if each person worked on their own to make an item from beginning to end. Tasks and equipment are shared and batches of products can be produced a number of times.

Batch production is a flexible method of production because similar products can be produced with only a small change in tooling. This allows the company to respond to changes in the market very quickly. Different templates and jigs are often used to respond to these market demands.

Batch production requires a number of people producing a limited amount of the same product

Mass production is an effective way of continously producing the same item

Mass production

This involves producing large numbers of the same product over a long period of time. There are two types of mass production; repetitive flow and continuous production.

Repetitive flow

The production is broken down into assembly and sub-assembly lines. Each person is trained to perform one specific job and the product or part is then passed on to the next person. This reduces time and cost in manufacture because training is only required in one area and machinery can be set to do one task repeatedly. Some assembly lines can be fully automated and require very little human input.

The drawback with this type of method is that if there is a problem or a machine breaks, the whole process may need to be halted. It also takes a lot of time and investment to change tooling so that another product can be made.

Continuous production

This involves the production of one item, 24 hours a day, for a continual period of time. Products such as steel, chemicals and food are produced in this way. This type of production continues because it would be costly to stop and then restart production. A small work force is usually needed to maintain the process.

Activities

1 Give three examples of products that are likely to be produced using one-off production methods.

2 List the advantages and disadvantages of batch production.

3 What are the differences between repetitive and continuous mass production methods?

Key point

● Different production methods are used depending upon the type of product being manufactured and the volume required.

Industrial applications 2

Manufacturing systems

Manufacturers often employ different systems in their method of production. Planning manufacture is often a complex process. It involves managing machines, processes and people so that the product can be produced efficiently and cost effectively.

Cell production

A group of people work within a production unit or cell to produce one component or a number of similar components. Other cells at the same time are also producing components that are collected and passed on. All of these components are then assembled together to produce the final product.

In-line assembly

This a traditional method used to produce many everyday items. The assembly takes place on a continuous assembly line with a separate component being added and then passed on for the next component to be added. This process continues until the product is completely assembled and finished.

Control

To ensure that all of the products are manufactured to the same quality, control checks take place during manufacture. The quality control team take samples of the product at different points along the production line. These are then examined and tested for faults or poor quality production against the required standard. This can therefore result in procedures taking place such as machines being repaired or operations being reset. This ensures that the consumer receives a safe and quality item.

In-line assembly

Just in time

This method relies heavily on logistics – the availability of materials and components when they are required. The system works on the fact that materials, components and sub-assemblies from other manufacturers arrive to be assembled into the final product immediately. The final assembled product is then also sent out as soon as it is completed. The reason this is done is:

- the manufacturer does not need a warehouse to store stock or the finished product

- the product can be changed quickly without using up the existing stock first.

The advantage is that the product can be changed according to demand without all of the company's money sitting in the unsold products.

Other factors

Once the product has been manufactured, it needs to be packaged and distributed. This is a very important part of the production if it is to be sold. The functions of the packaging are to:

- Protect the product so it is not damaged when transported and arrives at its destination in good condition. This is very important if the product has to be stored and transported across the world.

- Promote and advertise the product. How the package is designed has an influence on who it appeals to. The packaging helps in attracting the target market and is often produced in conjunction with the advertising and marketing of the product. A variety of packaging may be produced that appeals to people from different countries and cultures.

- Give information on how the product is to be used or maintained, and any safety requirements.

Toyota is credited with discovering 'just-in-time' management – an efficient system of stock control

Activities

1 What are the advantages and disadvantages of:
   - cell production?
   - in-line assembly?
   - just in time?

2 Why are packaging, distribution and marketing so important?

Key points

- Manufacturers will use systems in order to produce products in an efficient and cost effective way.

- Control procedures are put into place in order to maintain quality for the consumer.

- Packaging, distribution and marketing are important factors in the sale of the product.

Questions

1 Explain what a Gantt chart is used for.

2 What should a production schedule include?

3 Flow charts are often used to plan processes.

   a Draw the three basic symbols used in a flow chart.

   b Draw a flow chart with feedback to show the following processes:

      i vacuum forming

      ii line bending.

4 Health and safety is an important issue for the designer, manufacturer and customer.

   a When selecting materials to use, what health and safety issues does the designer have to consider?

   b List three health and safety requirements that the school workshop must conform to and explain why they are important.

   c What personal safety equipment should be used when operating a pillar drill?

   d What is risk assessment and why is it carried out for each activity?

   e Explain three ways a manufacturer can alter their working practices to have less impact on the environment.

5 Manufacturers use different production methods in order to manufacture their products. The type of product and the amount of units being produced will often determine which production method is used.

   a Which production method would be most appropriate to produce 800,000 units of the same product per week? Explain why this is the best production method to use.

   b Explain the difference between repetitive flow and continuous production.

   c Which production method would be the most appropriate to produce 1000 units of four different products per week? Explain why this is the best production method to use.

   d Manufacturers often employ different systems in their method of production. Explain the advantages a just in time system offers the manufacturer.

PRODUCT EVALUATION

Evaluation

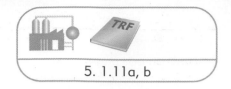
When do you carry out an evaluation?

The evaluation is not just something that occurs at the end of the project. The very best coursework has evaluations at different stages all the way through. There are two types of evaluation you will need to carry out when designing and making the product – on-going and final evaluations.

On-going evaluation

The on-going evaluation occurs when you make a decision or judgement about your work. The analysis and evaluation of your research will lead you to write a detailed specification. This specification therefore becomes the criteria with which you can make judgements when:

- evaluating your initial ideas, developed ideas and models in order to select the best design
- testing, assessing and evaluating the product as it is being made
- evaluating and assessing the time, materials and resources needed to make the product.

The specification is your guide to designing and making the product. In reviewing your work against the specification at regular intervals you are ensuring that a quality product is being produced that fulfils the design need.

Final evaluation

This is carried out when the product is finished to see how successful the product and production methods were. The product must be compared against:

- fitness for purpose
- the design need
- the needs of the intended user or users.

All products are evaluated to see how successful or unsuccessful they are

In order to do this a complete evaluation must be made. The final evaluation can be broken down into stages to simplify the process and to ensure that you carry out a thorough evaluation. The specification you wrote at the beginning of the project is your guide to see if the product is successful. The stages are:

- Evaluating the product against the specification.
- Carrying out tests on the product.
- Evaluating your own performance.
- Evaluating the production system (jig, pattern etc).
- Suggesting where modifications and improvements could be made.

Evaluating the product against the specification

For each point in the specification you wrote at the beginning of the project you will have to compare the product against it. For each point ask yourself 'Did I fulfil this point? Does the product do what I said it must do?'

Function and performance
- ❑ Does the product do what it was intended to do?
- ❑ Does it work in the environment it was intended to?

Size
- ❑ Is the size of the design appropriate?
- ❑ How does it fit any specified dimensions, e.g. anthropometrics, product sizes?

User
- ❑ How has the product been designed to appeal to the target market?

Maintenance
How have you ensured that the product:
- ❑ can be kept in good working order?
- ❑ can be easy to repair?
- ❑ can be easy to clean?

Life span
- ❑ How have you ensured that the product will last its intended life span?

Safety
What precautions have you taken to ensure that the product:
- ❑ is safe for the user?
- ❑ is safe to make?

Cost
- ❑ How have you managed to keep the cost low enough for your target market?

Production time
- ❑ Have you managed to complete the project by the required deadline?

Materials
- ❑ Why have you chosen these materials and components?
- ❑ Would any other materials and components be appropriate?
- ❑ How are these materials and components fit for their purpose?

Manufacture
- ❑ Have you chosen the appropriate methods for manufacture?
- ❑ Did these methods ensure that the product could be produced in quantity?
- ❑ Did these methods ensure a quality outcome?
- ❑ Did you make good use of the time, equipment and materials?

You will then have to give reasons and evidence as to why it did or did not fulfil each specification point. In carrying out tests, reviewing your performance and evaluating the production system, you will acquire the evidence that can be used to justify your reasons.

Evaluation checklist

The above checklist can be used as a guide to evaluating the product. This checklist links in with the section on writing the specification. Remember, you do not have to use all of these headings. Some of them may not be relevant to your product or the specification you have written.

Activities

1. What is the difference between an on-going and final evaluation?
2. Why is the specification used as the criteria for an evaluation?
3. In what ways can you gain evidence for your final evaluation?

Key points

- There are two types of evaluation – on-going and final evaluation.
- Evaluations use the specification as the criteria.
- When carrying out a final evaluation you must always give reasons for your answers with evidence to back them up.

Testing the product

To see how successful the product is you will have to carry out testing. In doing so you will be able to judge the product's strengths and weaknesses. The results of the tests you will carry out can be used as evidence in writing the evaluation. Testing will also highlight where improvements and modifications can be made or where further development is needed. There are three ways in which the product can be tested:

- by the user
- by an expert
- by the designer.

The user

What does the user think about the product?

Allowing the user to try out the product will give you an indication of whether it:

- fulfils its need
- functions correctly
- appeals to the user.

To get a good cross-section of opinion a number of different users should test the product. These users should be from your target market – the type of person you were aiming the product at. The testing of the product can be recorded in two ways:

- Observing the user interacting with the product and writing down what you have discovered. Taking photographs may also assist in the explanation of your observations.
- Asking the user to complete a questionnaire to obtain their thoughts and opinions.

The expert

You can also ask experts their opinions of a product. In doing this you will get a more in-depth feedback as the expert will have greater knowledge of the product. The questions can therefore be more technical or related to their area of expertise.

The designer

The designer and manufacturer in this case are you. Throughout the design process you have been making judgements. You may have carried out research and tests on materials, finishes, components etc. in order to make the correct selection. You will now have to test the product in order to see if the selection is fit for its purpose. The type of testing you will carry out will depend on the type of product. The type of tests may be:

- repeatedly using the product to see how reliable and durable it is
- placing weight on the product
- establishing how stable the product is
- placing the product in its intended environment.

When carrying out these types of test, try not to test the product to breaking point or ruin it. Also try to apply a method to how you carry out the test. For example if you were to test a drawer you could open and close it a set number of times to see how the runners and handle stand up to use. This could be carried out again with weight placed in to see if there is a difference.

What to test

All products should be tested for reliability, function and safety. Ask yourself, 'Is this product fit for its purpose and what tests can I carry out to see if is?' Using guidelines set out by the British Standards Institute will also be useful in deciding what to test.

Your specification is a guide for what to test. For example, if one of your specification points is, 'The product must be easy to operate' then ask the user to rate the ease of use on a scale of 1 to 5. For certain specification points such as, 'The product must have a durable finish,' you will be able to carry out the test yourself. This may be done by repeatedly using the product to see how well the finish wears.

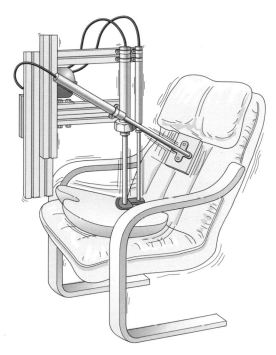

Test rig

What have you found?

After the testing is complete you will have to collate all of the results and display them.

- Quantitative questions you have asked in the questionnaire can be displayed as graphs, charts and tables.

- Answers to qualitative questions can be summarized.

- Observations or expert opinions can be written as a report.

In conclusion

It is important to remember that your final product is a working prototype. In carrying out tests, conclusions have to be drawn resulting in proposals of how the product could be modified. This is exactly what designers and manufacturers do when they develop their products for large-scale production. In writing the summary it is necessary write about:

- what worked well

- what did not work well

- what modifications and improvements are needed

- what further development is needed.

Activity

1 Why is the product tested by the:
- user?
- designer?
- expert?

Key points

- Testing allows you to determine how successful or unsuccessful your product is.

- Testing allows you to gain evidence to justify your reasons in evaluating against the specification.

- The product can be tested by the user, the designer and by an expert.

Review of your performance

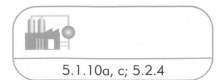

As well as evaluating the product, it is also important to evaluate how you have performed. This will allow you to review how you approached the project and the skills that you used. It will help explain to the person marking the project what problems you encountered and how you might approach it differently next time.

Setting it out

You will have used a wide range of skills to produce the product. When setting out to review your performance it will help you to break it down into sections. This will help you focus on a particular area. The two key areas are:

1 Planning

2 Making.

Planning

In preparing to make the product you will have produced a range of plans. This was done in order for you to organize your time and resources so that the product was completed on time with a quality outcome. Now that you have made the product, ask yourself, 'Were the plans effective for you to complete the product successfully?' Opposite is a checklist for you to review your performance in planning the making activity.

This should be used as a guide. Some of the headings may not apply to your planning but try to give reasons and justifications to all of the answers.

Activity	Lesson									
Time plan for the construction of a jewellery box **All lessons 2 hours** **Total time 20 hours**	1	2	3	4	5	6	7	8	9	10
Mark out sides, lap joints and rebates Cut out and check for fit	X	X								
Glue sides together Measure and mark lid/bottom Cut out and check for fit			X							
Set up CAD/CAM for lid Cut out detailing on lid Glue lid and bottom into rebates				X						
Clean up/sand box Mark where lid is to be cut Cut lid					X					
Mark out rebates for hinges Chisel out rebates					X	X				
Test fit hinges Mark out lock for box Cut/drill to fit lock							X			
Test fit lock Clean/sand inside of box							X			
Cut out velvet for the inside Test fit velvet								X		
Final clean up/sand inside and outside box Vanish box								X		
Sand box down lightly Apply second coat of varnish									X	
Glue in velvet Fit hinges and lock										X

Time plan

Time plan
❑ Did you manage to complete the product by the deadline? If not, what were the reasons for this?
❑ Was the time plan used effectively? Did you manage to complete each section by the required date?
❑ Was the time plan realistic? Was enough time given for each task?

Flow charts
❑ Did the flow chart(s) assist in controlling the outcome of the process?
❑ Was more detail needed? If so, what was needed?

Production schedule
❑ Did the production schedule have enough detail?
❑ Was it placed in the correct sequence? If not, how could it have been modified?
❑ Was there enough time for each stage to complete the task? Did you under-estimate the time at any stage?
❑ Did you follow the health and safety procedures set out for each task?
❑ Did you follow the production schedule? If not, what did you do differently and why?

Process plan
❑ Did the process plan(s) contain all of the information needed in order to carry out the process successfully? If not, what was needed to improve it?

Working drawing
❑ Did the working drawing contain enough information for you to produce the product? Could it have been improved in any way?

Your skills
❑ What difficulties did you come across when making the product? What did you find easy? What did you find hard?

Tools and equipment
❑ Did you select the correct tools and equipment for each task?
❑ Were the tools and equipment available adequate to construct the product?
❑ Could any other tools and equipment have been used that would have been better?
❑ What health and safety equipment did you use and was it adequate?

Materials and components
❑ Did you have any problems working with any of the materials? If so what problems did you encounter and what could have been done to prevent these problems?
❑ Did you have any difficulty with standard components, fixtures and fittings? If so what problems did you encounter and what could have been done to prevent these problems?

Making

During the making stage of the project you will have used the working drawing and plans to assist you. In constructing the product you will have:

● drawn upon your own experience and skills

● used a range of materials and components

● used a range of tools and equipment.

The checklist opposite can be used as a guide to evaluate your performance in making.

Activities

1 Why is planning so important?

2 What three areas do you need to evaluate in making?

Key points

● Your performance can be reviewed in your planning and making. It also allows you to gain evidence to justify your reasons in evaluating against the specification.

● Evaluating your performance allows you and others to see how you approached the project, the skills that you used and what difficulties you came across.

Production systems

In making the product you will have used one or more of the following:

- jig
- template
- pattern
- mould
- computer system

These production systems allow the product to be manufactured in quantity. They also allow the control of accuracy to ensure consistency and quality.

Evaluating this part of the production will allow you to see how successful or unsuccessful the production system was in producing the product. You will then be able to suggest ways it could be improved and developed in order for the product to be manufactured for sale.

Evaluating the control device

Having used the control device in the production of the product, you will have to ask yourself, 'How successful was it?'

Use the following checklist as a guide to evaluate the control device. Due to the fact that there are a number of different types of control device that could have been used, the checklist tries to be generic to allow the evaluation of them all. Therefore you may want to add comments that are specific to the particular device that you have used. When using the checklist try to give reasons, justifications and evidence in the answers that you give.

Design and function
- ❑ Did the design of the device allow it to be used:
  - easily?
  - quickly?
  - repeatedly?
  - accurately?
- ❑ What improvements and modifications could be made to the design in order for it to function better?

Production
- ❑ Were the correct tools, equipment and materials used for the production of the device?
- ❑ Can it be used repeatedly and accurately?
- ❑ Was the production of the device cost effective?

- ❑ How long did it take to produce and was this worth the time invested into producing the device?
- ❑ What testing took place and what modifications were made so that it functioned correctly?
- ❑ What improvements could be made to it if you were to produce it again?

Industrial applications

In the section on industrial applications we dealt with how manufacturers used different production methods and systems in order to produce their goods. Designers often produce prototypes, much in the same way that you have done, before manufacture begins.

Having produced a prototype using the control device, you will need to ask yourself the question, 'How could my design and control device be developed in order for it to be manufactured for sale?' In order to do this you will have to state:

- what particular production method would be most suitable for your product and why

- what changes would have to be made to the prototype and the control device in order for it to be manufactured efficiently and why.

Finally you will need to briefly describe how you would plan the production of your product.

It may be useful to use diagrams or flow charts to help explain the process of manufacture. Try to include:

- the organization of the workforce

- the layout of the workshop or factory

- tools and equipment needed

- the storage of materials and of the final product.

Activity

1 Why is a control device important for manufacturing products?

Key point

- Control devices allow the product to be made in quantity and ensure consistency and quality. Evaluating the control device allows you to see whether it fulfilled its function and how it can be used to produce the product for manufacture.

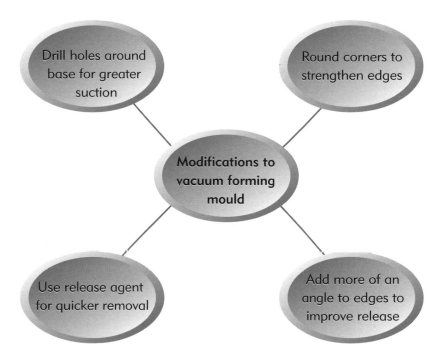

Drill holes around base for greater suction

Round corners to strengthen edges

Modifications to vacuum forming mould

Use release agent for quicker removal

Add more of an angle to edges to improve release

Modifications and improvements

5.1.11c–e; 5.1.10c; 5.2.4

Modifications during making

While you were making the product you may have made modifications or changes to it. This may have been done for several reasons, such as:

- there was a better way of making the product

- to improve the product's function or appearance

- to rectify mistakes.

Look back at your plans, working drawing and how the product was constructed. Write down any changes, modifications, alterations and improvements you have made. Always give reasons as to why this was done to justify your actions.

The final evaluation

In this section you have carried out a thorough evaluation of the product by comparing it to the specification and using evidence to justify your answers from:

- testing

- evaluating the control device

- evaluating your performance.

After carrying out your final evaluation, what conclusions have you made? For each of the areas above you will have written down where improvements, modifications and developments could have been made. These will form the basis for writing the final part of the evaluation.

Along with this there may be other areas where modifications and improvements could be made to the product. When carrying out the evaluation against the specification you may have found that certain points were not quite fulfilled.

The final evaluation will offer you the opportunity to propose where modifications can be made to the product and to how it was made. Ask yourself the following questions:

1 What changes and modifications do I need to make to the product in order for it to:

- function better?

- fulfil the needs of the user better?

2 If I were to make the exact same product again, what would I do differently with regards to:

- the materials and components used?

- how it was constructed?

Other issues

In examining the way your product could be improved or modified you will have to consider moral, social, cultural and environmental issues. These issues are important when designing a product to be manufactured. Ask yourself the following questions:

- Can other materials and components be used to make the product more environmentally friendly?

- Can the product be re-designed to minimize the use of materials?

- Would the product only appeal to people from particular cultures? Can the product be altered to appeal to a wider group of people with different cultural backgrounds?

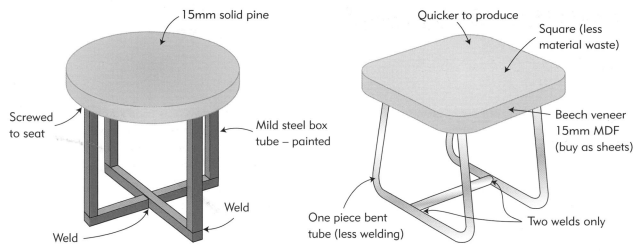

Stool before

Stool after

● Does the product have any impact on social issues such as isolating people or making them feel vulnerable in any way? Can it be modified so that it brings people together or includes more people?

● Does the product impact on any moral issues such as the materials and components being manufactured by people with bad working conditions and low wages? Does the product have the capacity to cause harm indirectly? Can it be modified to prevent this?

Development of the product

Look back to the industrial applications section of your evaluation. In this section you will have written about how the product can be modified in order for it to be manufactured. You have also identified where the product can be modified and improved in this section of the evaluation. Using these as your guide you will now have to offer ways in which your product can be developed in order for it to be manufactured. This can be done through re-designing the product using sketches and annotation to show the different possibilities for the product's development.

In carrying out this development there are extra considerations to take into account. In the design try to consider:

● the cost of producing the product in quantity

● how long it might last without breaking or wearing out

● how often it needs to be maintained

● how easy it is to maintain.

Activities

1 What reasons are there for modifying the product during manufacture?

2 Why is the product developed at the end of the evaluation?

Key points

● Modifications can occur during the making of the product to ensure a better outcome.

● The evaluation process highlights where improvements can be made.

● Modifications can be suggested after carrying out the evaluation process to improve the product.

● The modifications suggested in the evaluation can be used as the criteria in which to further develop the product for manufacture.

Questions

1 There are two types of evaluation that can be carried out in a design folder, on-going and final evaluation. Explain the difference between the two.

2 When carrying out an evaluation, what is used as the criteria?

3 Testing the prototype product is an essential part of the evaluation. It will help identify the product's strengths and weaknesses.

   a Why should a cross-section of users from your target market test the product? Explain your answer.

   b Suggest two ways in which you can record the user testing the product.

   c What benefits would there be in asking an expert to test the product?

   d Describe in detail a test that could be carried out on a TV cabinet. Use diagrams to help explain your answer.

4 Describe how you would carry out an evaluation of your performance in planning and making.

5 The control device is used to manufacture a product in quantity. Describe what other benefits a control device can offer.

6 How can you judge whether the control device was successful or not? Explain your answers.

7 When a prototype is being produced, modifications may be made to it. Explain why this happens.

8 At the end of the evaluation, modifications and improvements to the product are suggested. Explain why this is beneficial.

9 *Re-developing the product after the evaluation has taken place can lower the manufacturing cost.*

   a Give reasons why you would or would not agree with this statement.

   b What other benefits can redevelopment give?

MAKING THE GRADE

Presentation

Why presentation is important

Marks are awarded for the content of your folder and how it is presented. Your folder work represents you and your understanding of the subject. In presenting your folder work neatly you are making an overall visual impact and demonstrating that you care about your work.

Borders

To give an overall identity to your folder it is a good idea to have the same border on every page. This gives continuity to your folder work. Remember that highly decorative borders do not gain you extra marks. In fact it can take up valuable time if you are re-drawing and colouring a border for every page. Try to draw, by hand or on the computer, a simple border that includes your name. You may also want to add on a title or personalize it with a logo that you have designed. Your border can then be photocopied for the required amount of pages.

Quality vs quantity

Sheets in your folder that are full of information and detail look much better than sheets which are empty. Producing 70 A3 sheets does not mean you will get more marks than someone who has produced 30. It is recommended that the whole project should be between 25–30 A3 pages in length. This means that each page should be filled with quality information.

Consistency

To give a good overall appearance it is important to be consistent. Along with the border, there are other things that you can do in order to enhance your folder's appearance. The following is a list to help you in the presentation of your work:

- If you are word processing work for your folder make sure to use the same font and font size throughout the folder. Try to do the same for headings and sub-headings.
- Use the same writing style for headings and sub-headings.
- Write text in black pen and annotations on sketches in pencil.
- Limit the number of colours you use.
- Don't start to draw headings in a fancy style that takes a long time to do.
- If you are unsure of spellings or content, write it in pencil first. This allows corrections to be made before you write over in pen.

Finishing touches

When the folder is completed you may have some spare time. In this time you can add finishing touches to complete the folder's presentation. This will also assist the examiner in marking the folder.

- Put the pages into order and number them.
- Write a contents page to put at the beginning.
- Draw a front cover.
- Bind the folder together.

Objective checklist

Throughout this book you have been encouraged to complete all of the necessary areas in order to complete the objectives. Your teacher will also guide you through the production of the folder and the making of the final product. The following pages aim to assist in the production of the coursework in offering suggestions on how to present the folder and break down each objective to give you a clear insight to what is required. In order to gain a good mark in the coursework the following guide has been produced:

Objective 1

The design brief you produce must be for a product that is marketable and can be produced in batches to be sold.

You must ensure that you have investigated the user and the consumer. In producing a consumer profile you will find it much easier because you will know exactly who you are designing for.

Objective 2

The production of the specification is one of the most important sections of the folder. You will be constantly referring to this all the way through the folder. Therefore the specification is just that – specific. Don't write vague points, which cannot be justified or proven like, 'It must be nice' or 'It must look good'.

Objective 3

You must present your ideas using a range of techniques. This must include ICT and CAD.

Objective 4

You must plan the construction of your chosen idea and select the appropriate tools, equipment, materials and methods. Modifications and development of your idea can take place as a result of carrying out tests and investigations.

You will have to design and develop the control device. This is to be made at this point so it can be used in the production of the product in Objective 5 and evaluated in Objective 6.

You must include CAD/CAM in this section to gain higher marks.

Objective 5

Remember 52 marks are available in this objective, more than the other five objectives put together. It is extremely important that you get your planning and making right in order to achieve a high final grade. You must produce detailed plans in order to gain higher marks. This also allows you to see exactly what you are doing when you are making the product and as a result achieve higher marks.

The size of the product you make needs to reflect the time available. You may want to make a smaller, detailed item that can be completed in the time available.

Objective 6

To gain high marks in this section you must carry out a complete evaluation. This is done in order for you to gain evidence so that you can justify your reasons and opinions. The conclusions you make from carrying out the evaluation will inform where development, improvements and modifications can take place.

Communication

It is important that your folder communicates what you have done and why you have done it. Your folder work must therefore be logical and the different sections must relate to one another. You must be able to go through the folder and see that it is a continual, flowing process from start to finish, not separate sections. This can be achieved through the presentation of your work and the quality of your written work to make sure that you are clearly understood.

Internal assessment: Objective 1 (Total marks 4)

Identification of a need or opportunity leading to a design brief

Candidates will need to:

- provide a description of the design need using various means of communication

- identify the range of users and the market for which the product is intended

- develop a design brief for a marketable product.

In order to achieve the full marks available for this section you will need to provide proof that you have:

- provided a description of the design need using various means of communication such as:

  – writing

  – using images and graphics – graphs, photographs, clippings etc.

- identified the range of users and the market for which the product is intended. This can be done through analysing the user and the market.

User

- Questioning the user, through a survey, about their opinions and needs.

- Analyzing the lifestyle of the user to provide a consumer profile.

Market

- Examining what products are available to buy.

- Developing a design brief for a marketable product using what you have found out in your investigation to identify the need. Writing a clear brief based on the need that is aimed at producing a marketable product (a product which can be made and sold commercially).

Statement

A new gallery called 'Far away places' is due to open before Christmas 2002. I have been asked to design and make a contemporary clock, decorative box, or item of jewellery to add to their extensive gift range. The item I make must be influenced in some way by a culture other than my own.

Throughout my major project I need to take into account these key factors:
✓ Quality control (a quality control device must be made)
✓ Originality
✓ Uniform to the core of the culture
✓ Continuity of my theme throughout the project

Situation/need

My jewellery box will be designed to satisfy the need for a box that will contain all of one's jewellery without being too large or aesthetically displeasing. The box will allow people to place all their jewellery into a good-looking culturally influenced box, knowing that their jewellery is tidy stored properly and in one place.

Design brief

I wish to make a jewellery box influenced by *oriental* culture. I have chosen to influence my project by the oriental culture because it intrigues me and I believe it has a great potential with the product I intend to make. I feel that the various symbols and letters of many oriental alphabets are attractive because they are so refreshingly different to the majority of western alphabets, they are also aesthetically pleasing because of their various curves and varying thickness of lines.

The reasons for my decision to design and make a jewellery box are:
I imagined from the start that a modern jewellery box with ancient traditional roots would be easier for me to design than a clock or a piece of jewellery.
The product will be useful to my user group and will be affordable.
The jewellery box may be used for a long time, because it will be durable.
The materials that I choose will be modern (from a financial point of view) but the materials I will use that will be visible will hopefully be natural and from the Far East to add authenticity to my box. The box will be a clever mix of cheap modern materials that are strong and resilient, and natural material from the Far East that will help provide the design and cultural influence to my design.

My product must meet the following targets:
✓ Well-designed original product
✓ Good manufacture and finishing
✓ Attractive to the target group
✓ Not too expensive to manufacture
✓ Durability and practicality.

User Group

My target consumers will be people who are in almost any age category, but interested in oriental culture. The jewellery box must be practical because although my user group is interested in oriental culture they will not want to compromise practicality for style, they will want both.

The user group may have many items of jewellery, which they would want to keep in one place and not in several smaller boxes. This jewellery box therefore needs to be substantial in size but not so big that it intrudes on its surroundings.

A student's identification of an opportunity, and design brief

Internal assessment: Objective 2 (Total marks 12)

Research into the design brief which results in a specification

Candidates will need to:

- examine the intended purpose, form and function of the product

- undertake appropriate surveys, identifying and evaluating how existing products fulfil the needs of their intended users

- identify and collect data relevant to the product(s) and its users

- develop a detailed specification and criteria that includes the capability for batch production.

In order to achieve the full marks available for this section you will need to provide proof that you have:

- examined the intended purpose, form and function of the product by:

  – examining and evaluating existing products on the market against a criteria (function, aesthetics, fitness for purpose etc.)

  – examining and evaluating existing products against a criteria by using them and disassembling them

- undertaken appropriate surveys, identifying and evaluating how existing products fulfil the needs of their intended users by:

  – producing and carrying out a user survey to gain opinions on existing products and how these products meet the user's needs

  – displaying the results in graphs

- identified and collected data relevant to the product(s) and its users by gathering data from a variety of sources in order to write a specification and to design the product. Examples include anthropometric data, standards, product sizes, batch production etc.

- developed a detailed specification and criteria that includes the capability for batch production by analysing the research you have collected to write a conclusion

- written a detailed design specification as a result of the analysis that provides a system so the product can be produced in batches.

Examples

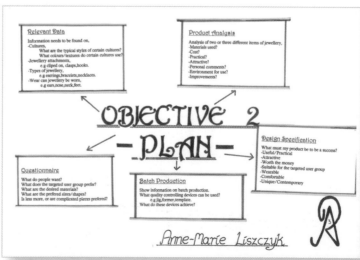

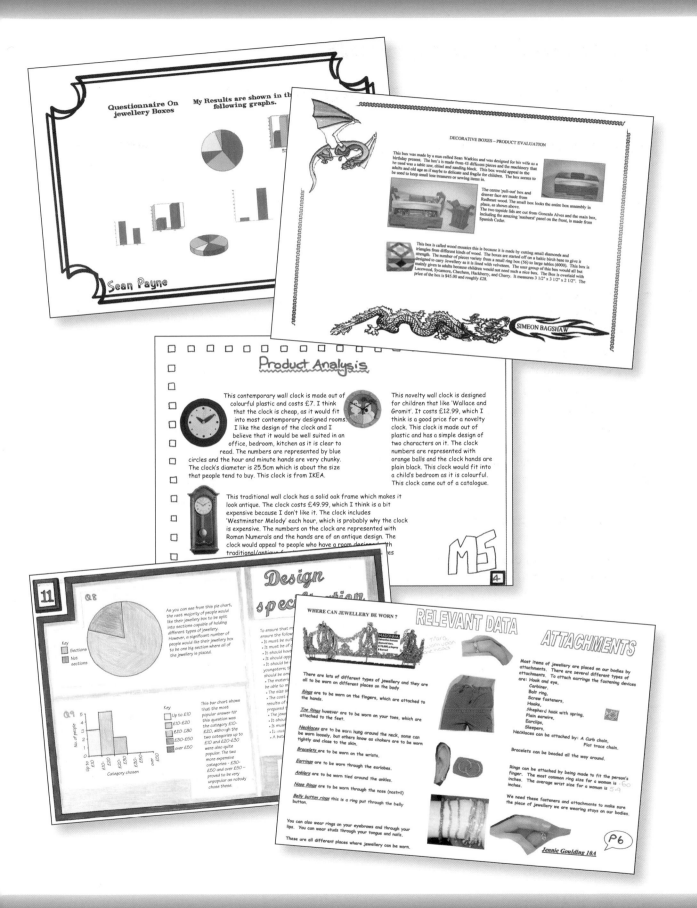

Questionnaire On jewellery Boxes

My Results are shown in the following graphs.

Sean Payne

DECORATIVE BOXES – PRODUCT EVALUATION

This box was made by a man called Sean Watkins and was designed for his wife as a birthday present. The box's is made from 45 different pieces and the machinery that he used was a table saw, chisel and sanding block. This box would appeal to the adults and old age as if maybe to delicate and fragile for children. The box seems to be used to keep small lose treasures or sewing items in.

The centre 'pull-out' box and drawer face are made from Redheart wood. The small box locks the entire box assembly in place, as shown above.
The two topside lids are cut from Goncalo Alves and the main box, including the amazing 'sunburst' panel on the front, is made from Spanish Cedar.

This box is called wood mosaics this is because it is made by cutting small diamonds and triangles from different kinds of wood. The boxes are started off on a baltic birch base to give it strength. The number of pieces variety from a small ring box (56) to large tables (6000). This box is designed to carry Jewellery as it is lined with velveteen. The user group of this box would all but mainly given to adults because children would not need such a nice box. The Box is overlaid with Lacewood, Sycamore, Chechem, Huckberry, and Cherry. It measures 3 1/2" x 3 1/2" x 2 1/2". The price of the box is $45.00 and roughly £28.

SIMEON BAGSHAW

Product Analysis

This contemporary wall clock is made out of colourful plastic and costs £7. I think that the clock is cheap, as it would fit into most contemporary designed rooms. I like the design of the clock and I believe that it would be well suited in an office, bedroom, kitchen as it is clear to read. The numbers are represented by blue circles and the hour and minute hands are very chunky. The clock's diameter is 25.5cm which is about the size that people tend to buy. This clock is from IKEA.

This novelty wall clock is designed for children that like 'Wallace and Gromit'. It costs £12.99, which I think is a good price for a novelty clock. This clock is made out of plastic and has a simple design of two characters on it. The clock numbers are represented with orange balls and the clock hands are plain black. This clock would fit into a child's bedroom as it is colourful. This clock came out of a catalogue.

This traditional wall clock has a solid oak frame which makes it look antique. The clock costs £49.99, which I think is a bit expensive because I don't like it. The clock includes 'Westminster Melody' each hour, which is probably why the clock is expensive. The numbers on the clock are represented with Roman Numerals and the hands are of an antique design. The clock would appeal to people who have a room designed with traditional/antique furniture.

MS

4

11

Q8

As you can see from this pie chart, the vast majority of people would like their jewellery box to be split into sections capable of holding different types of jewellery. However, a significant number of people would like their jewellery box to be one big section where all of the jewellery is placed.

Key
Sections
Not sections

Q9

This bar chart shows that the most popular answer for this question was the category £10-£20, although the two categories up to £10 and £20-£30 were also quite popular. The two more expensive categories – £30-£50 and over £50 – proved to be very unpopular as nobody chose these.

Key
Up to £10
£10-£20
£20-£30
£30-£50
over £50

No. of people

Category chosen

Design specification

To ensure that my ... ensure the follow...
• It must be suit...
• It must be of c...
• It should have...
• It should appe...
youngsters; th...
should be sm...
• The materia...
be able to m...
• The size o...
• The cost...
results of...
prepared t...
• The jewel...
• It shou...
• It mus...
• It use...
• A batt...

WHERE CAN JEWELLERY BE WORN ?

RELEVANT DATA

MADONNA

Tiara worn upon head

There are lots of different types of jewellery and they are all to be worn on different places on the body

Rings are to be worn on the fingers, which are attached to the hands.

Toe Rings however are to be worn on your toes, which are attached to the feet.

Necklaces are to be worn hung around the neck, some can be worn loosely, but others know as chokers are to be worn tightly and close to the skin.

Bracelets are to be worn on the wrists.

Earrings are to be worn through the earlobes.

Anklets are to be worn tied around the ankles.

Nose Rings are to be worn through the nose (nostril).

Belly button rings this is a ring put through the belly button.

You can also wear rings on your eyebrows and through your lips. You can wear studs through your tongue and nails.

These are all different places where jewellery can be worn.

ATTACHMENTS

Most items of jewellery are placed on our bodies by attachments. There are several different types of attachments. To attach earrings the fastening devices are: Hook and eye,
    Carbiner,
    Bolt ring,
    Screw fasteners,
    Hooks,
    Shepherd' hook with spring,
    Plain earwire,
    Earclips,
    Sleepers.

Necklaces can be attached by: A Curb chain,
    Flat trace chain.

Bracelets can be beaded all the way around.

Rings can be attached by being made to fit the person's finger. The most common ring size for a woman is -60 inches. The average wrist size for a woman is -5.4 inches.

We need these fasteners and attachments to make sure the piece of jewellery we are wearing stays on our bodies.

P6

Jennie Goulding 10A

Internal assessment: Objective 3 (Total marks 12)

Generation of design proposals

Candidates will need to:

- generate a range of design proposals

- check design proposals against the design specification and review and modify them if necessary

- identify a chosen design proposal for product development

- present design solutions using a range of graphic techniques and ICT including computer-aided design (CAD), to generate, develop, model and communicate design proposals.

In order to achieve the full marks available for this section you will need to provide proof that you have:

- generated a range of design proposals by:

  - sketching a range (three or more) of different, initial ideas that are appropriate

  - developing selected designs through sketching and modelling

- checked design proposals against the design specification and reviewed and modified them if necessary by carrying out an on-going evaluation, against your specification

- identified a chosen design proposal for product development by:

  - carrying out an evaluation, against the specification, of your initial ideas and developed ideas

  - selecting suitable design ideas that can be developed further

- presented design solutions using a range of graphic techniques and ICT including computer-aided design (CAD), to generate, develop, model and communicate design proposals.

Examples

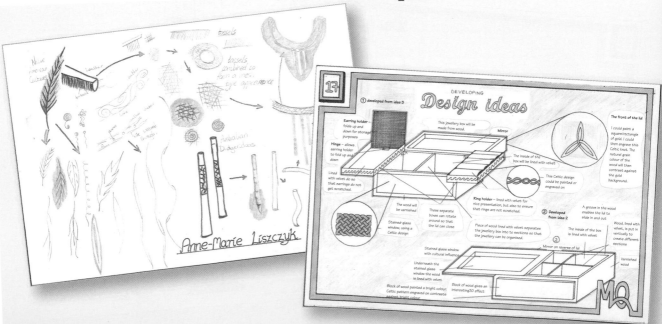

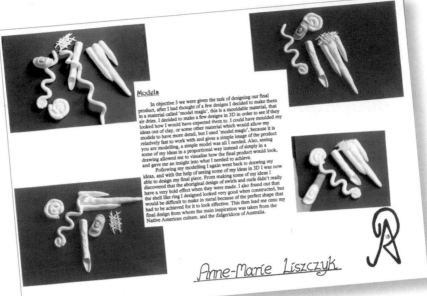

Models

In objective 3 we were given the task of designing our final product, after I had thought of a few designs I decided to make them in a material called 'model magic', this is a mouldable material, that air dries. I decided to make a few designs in 3D in order to see if they looked how I would have expected them to. I could have moulded my ideas out of clay, or some other material which would allow my models to have more detail, but I used 'model magic', because it is relatively fast to work with and gives a simple image of the product you are modelling, a simple model was all I needed. Also, seeing some of my ideas in a proportional way instead of simply in a drawing allowed me to visualise how the final product would look, and gave me an insight into what I needed to achieve.

Following my modelling I again went back to drawing my ideas, and with the help of seeing some of my ideas in 3D I was now able to design my final piece. From making some of my ideas I discovered that the aboriginal design of swirls and curls didn't really have a very bold effect when they were made. I also found out that the shell like ring I designed looked very good when constructed, but would be difficult to make in metal because of the perfect shape that had to be achieved for it to look effective. This then lead me onto my final design from whom the main inspiration was taken from the Native American culture, and the didgeridoos of Australia.

Anne-Marie Liszczyk

IDEAS

My idea for this box was to divide it as shown above, leaving one quarter as a metal plate with holes to secure earrings in, which would rotate to reveal another compartment.

3 Louisa Atkin

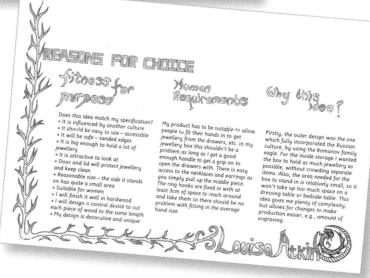

REASONS FOR CHOICE

fitness for purpose

Does this idea match my specification?
• It is influenced by another culture
• It should be easy to use – accessible
• It will be safe – sanded edges
• It is big enough to hold a lot of jewellery
• It is attractive to look at
• Door and lid will protect jewellery and keep clean
• Reasonable size – the side it stands on has quite a small area
• Suitable for women
• I will finish it well in hardwood
• I will design a control device to cut each piece of wood to the same length
• My design is decorative and unique

Human Requirements

My product has to be suitable to allow people to fit their hands in to get jewellery from the drawers, etc. in my jewellery box this shouldn't be a problem as long as I get a good enough handle to get a grip on to open the drawers with. There is easy access to the necklaces and earrings as you simply pull up the middle piece. The ring hooks are fixed in with at least 3cm of space to reach around and take them so there should be no problem with fitting in the average hand size.

Why this idea?

Firstly, the outer design was the one which fully incorporated the Russian culture, by using the Romanov family eagle. For the inside storage I wanted the box to hold as much jewellery as possible, without crowding separate items. Also, the area needed for the box to stand in is relatively small, so it won't take up too much space on a dressing table or bedside table. This idea gives me plenty of complexity, but allows for changes to make production easier, e.g., amount of engraving.

3 Louisa Atkin

Internal assessment: Objective 4 (Total marks 12)

Product development

Candidates will need to:

- make reasoned decisions about:
  - materials
  - production methods
  - pre-manufactured standard components.

This is done by:

- considering how materials are prepared for manufacture and how pre-manufactured standard components are used

- modelling, to ensure the product meets the original design brief and its fitness for purpose

- considering the implications for quantity manufacture of:
  - materials and components
  - tools, equipment and processes
  - critical dimensions and tolerances

- developing a control system to be used in the manufacture of the product

- being flexible and adaptable in responding to changing circumstances and new opportunities

- making any necessary modifications to the chosen design

- giving details of the final design including a final product specification

- presenting design solutions using a range of graphic techniques and ICT including computer-aided design (CAD) to develop, model and communicate design proposals.

In order to achieve the full marks available for this section you will need to provide proof that you have:

- carried out investigations, tests or trials, to make reasoned decisions about:
  - materials
  - production methods
  - pre-manufactured standard components

by carrying out investigations and testing your developed chosen idea. This will lead you to making decisions about the most appropriate materials, production methods and components that can be used to make your product. You must produce sheets detailing the results of testing and justify your selection

- considered how materials are prepared for manufacture and how pre-manufactured standard components are used by:
  - developing your design idea to allow for the preparation of material and the use of pre-manufactured components
  - highlighting these features in the on-going evaluation

- modelled your ideas and applied test procedures ensuring the product meets the original design brief and its fitness for purpose by:
  - producing a series of models to help develop and refine ideas
  - testing the models and evaluating them against the specification

- considered when developing the product, the implications for quantity manufacture of:
  - materials and components
  - tools, equipment and processes
  - critical dimensions and tolerances by showing consideration in the development sketches, models, evaluations and testing; and in the production of a manufacturing specification

- developed a control system to be used in the manufacture of the product by:
  - producing sketch sheets and models showing how the control device can be produced
  - producing the final control device to be used in making the product

- been flexible and adaptable in responding to changing circumstances and new opportunities
- made any necessary modifications to the chosen design by:
  - re-designing and developing the idea and control device as a result of testing and evaluation
  - producing a modifications sheet that details the changes made to the chosen idea with justifiable reasons

- given details of the final design including a final product specification by:
  - producing a final idea sheet that shows exactly what is to be produced and concludes the development process
  - producing a working drawing for the construction of the product

- presented design solutions using a range of graphic techniques and ICT including computer-aided design (CAD), to develop, model and communicate design proposals.

Examples

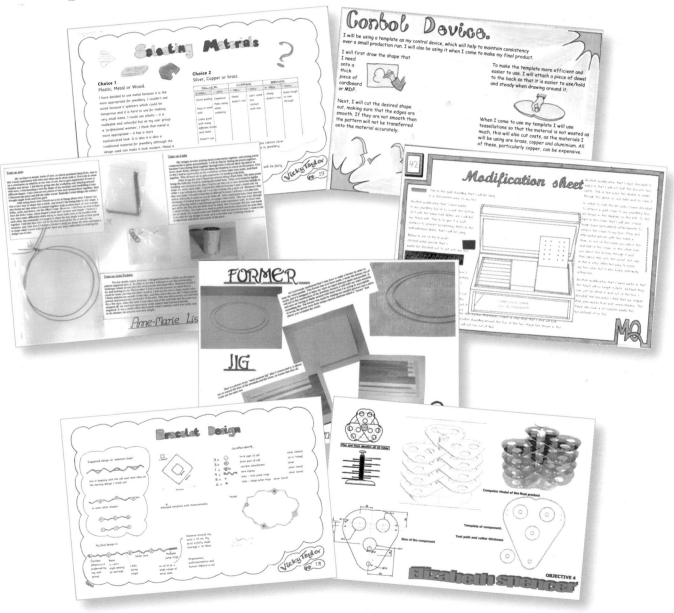

Internal assessment: Objective 5 (Total marks 52)

Product planning and realization

Candidates will need to:

- produce a plan of action which considers: materials, pre-manufactured items, equipment, processes and health and safety issues against an order of work and the need to make products that match the design specification

- select and use tools, equipment and processes effectively and safely

- economically prepare materials/pre-manufactured items for production, allowing for waste and fine finish

- complete a quality outcome suitable for the intended user or users, and ensure that their outcome functions effectively

- be prepared to adapt working procedures in response to changing circumstances

- use a range of skills and techniques appropriate to the task

- where appropriate apply a range of industrial techniques when working with familiar materials and processes.

In order to achieve the full marks available for this section you will need to provide proof that you have:

- produced a plan of action which considers: materials, pre-manufactured items, equipment, processes and health and safety issues against an order of work and the need to make products that match the design specification by producing a time plan, flow chart, production schedule and process plan

- selected and used tools, equipment and processes effectively and safely by:

  - following the production schedule and process plan

  - independently manufacturing the product safely using the appropriate tools, equipment and processes

- economically prepared materials/pre-manufactured items for production, allowing for waste and fine finish by being resourceful and adaptable with materials to achieve an accurate finish

- completed a quality outcome suitable for the intended user or users, ensuring that the outcome functions effectively by:

  - completing the product on time to a high standard that functions correctly and meets the final product specification

  - modifying the product and control device in order for it to function correctly

  - using quality control measures written into the plans

- been prepared to adapt working procedures in response to changing circumstances by:

  - testing and checking the product as it is being made

  - modifying the way the product is made in order for it to function correctly

  - use of alternative procedures in planning

- used a range of skills and techniques appropriate to the task

- applied, where appropriate, a range of industrial techniques when working with familiar materials and processes by:

  - using a range of different skills, processes and techniques to make the product which is appropriate to the design

  - the product having a high degree of precision through the use of the control device.

Examples

rhi

Flow chart for cutting figures

START

Mark out large main figure on aluminium and cut out using fretsaw

Files edges to a smooth finish

Mark out and cut the smaller figure, file edges

FINISH

FLOW CHART FOR ACID ETCHING

START

Are the edges smooth? — NO

Guillotine backplate of copper and wet-and-dry edges

Cover backplate in enamel and leave to dry

Mark out numbers on enamel with a scriber

Is enamel dry? — NO / YES

Place backplate in acid and leave for a couple of days

YES

Take backplate from acid and cover in paintstripper — NO — Is t ena comin

Use a toothbrush to remove enamel

FINISH

YES

FLOW CHART FOR FINISHING OFF

START

Cut and file the stand for the back of the clock

Mark mosaic pattern onto backplate

Silver solder the stand to the back of the clock

Buff front of backplate on buffing wheel

Wet-and-dry, and polish figures, polish backplate

Attach figures to backplate, attach clock mechanism

FINISH

CONSTRUCTION

START

Measure and mark out teak (pencil, tri-square, marking knife)

Measure and mark out inner box(pencil, tri-square, pencil)

Cut out sections of box with bansaw

Take the lid and cut down the sides (slanted/plain edges and sand down edges)

When finished place lid on CAD/CAM and leave to engrave design on

Cut out the lapped joints and dry fit them (bansaw)

Mark and cut out wood for hidden lock (pencil, marking rule, clamp saw)

Cut the front pannel in half (bansaw)

Chisel out hollow back of front pannel into 1/2 pin in (chisel, mallet)

Dry fit lid frame work and make sure lapped joints are aligned with box joints

Fix the wood for hidden lock on the back bottom pannel (nails, Hammer)

Put nail in the front of frame work. Make sure back and frame work fit (nail, Hammer)

fix lid down with glue to the framework

Mark out and cut out slots for box and include dividers(pencil, marking knife, saw)

Do more joints for skin and dry fit in place (extra saw)

Fit all the lapped joints together and put (front pannel on as well(glue,clamp)

Plain down the bootom then sand and smooth edges (downslanter, sander)

mark out groove for hinge at the back (pencil (chisel)

Fix hinge down lid and back pannel(screws screw drivers)

Fix divider down with groove (wood glue)

Apply green carpet to the bottom and sides of skin

Finishing touches add varnish to box if necessary

FINISH

SIMEON BAGSHAW

DESIGN PROCESS CHART		
PROCESS	EQUIPMENT	COMPLETE
Mark out sides, lid and bottom of box	Pencil, tri-square, steel ruler	
Cut out side's lids and bottom	Ban saw	
Mark out lapped joints	Pencil, marking gauge, tri-square, knife	
Mark out framework for lid	Tri-square, pencil	
Cut out framework on front and back panels	Hard back saw , ban saw	
Mark out lapped joints for the lid framework so it corresponds with box lapped joints	Pencil, marking gauge, tri-square, knife	
Cut and put joints together (dry fit) make sure they look neat and correspond	Hard back saw, ban saw	
Mark a horizontal line through middle of front panel	Pencil, tri-square	
Cut horizontal line through middle of front panel	Ban saw	
Dry fit again to see how much was lost during cut and plain down or sand any uneven parts	Sanding wheel, plain	
Setup CAD/CAM machine	CAD/CAM machine	
Leave CAD/CAM machine to do design on lid of box	Pencil, tri-square	
Mark out cutting section needed for the bottom of the front panel for lock		
Mark out the wood for lock	Pencil, cardboard template	
Cut out the inside of the bottom of front panel as marked in 14	Hard back saw or ban saw	
Put suitable nail into middle of front panel framework	Nail, pin hammer	
Mark out the lock for the lid to the box	Pencil, ruler	
Cut out locking section	Cutting saw, filpy	
Glue and fix framework for lid together don't stick lid on yet	Glue, G-clamps	
Nail on the lock to bottom of front panel in correspondence with nail on the lid	Pin hammer, nail	
Mark out wood for the skin of the box		
Cut out the skin of the box and glue together	Pencil, ruler, tri-square	
Ensure that the skin slides into box smoothly	Ban-saw, wood glue, 90degree clamps	
Mark out the velvet for bottom and sides	Pencil, ruler	
Glue lid onto framework	Wood glue, G-clamps	
Put plane brass on the back	Chisel, screws, screw driver	
Glue of stick down velvet	Glue	
Finishing touches wire wood polish or wax and remove any glue left round edges	Any necessary finishing equipment mentioned or used	
FINISHED		

SIMEON BAGSHAW

Internal assessment: Objective 6 (Total marks 8)

Evaluation and testing

Candidates will need to:

- evaluate their products to ensure that they are of a suitable quality for the intended users
- carry out testing, resulting in reasoned conclusions that suggest any necessary modifications to improve the product
- review whether they have used resources appropriately, e.g. time, materials, equipment and production methods
- analyze the performance of their manufacturing control system in the production of the prototype.

In order to achieve the full marks available for this section you will need to provide proof that you have:

- evaluated the products to ensure that they are of a suitable quality for intended users by:
  - evaluating the product against the specification using evidence to justify your answers
  - allowing the users to test the product and recording their opinions
  - carrying out further quality tests on the product

- carried out testing, resulting in reasoned conclusions that suggest any necessary modifications to improve the product by:
  - carrying out tests on the product
  - producing a modification sheet detailing where improvements can be made and how the product can be developed
- reviewed whether you have used resources appropriately, e.g. time, materials, equipment and production methods by:
  - carrying out a review of your performance in planning and making
  - testing the product
  - producing a modification sheet detailing where improvements can be made and how the product can be developed
- analysed the performance of your manufacturing control system in the production of the prototype by:
  - evaluating the control device and manufacturing system used and suggesting where improvements and modifications could take place
  - suggesting ways in which the control device and system could be improved, modified and developed in order to manufacture a saleable product.

Examples

Evaluation

At the start of this project, I set out to design and make a trinket box inspired by a foreign culture which was for an adult age group, and in my view I have succeeded in doing that.

My box meets most of the criteria of the design specification, however it doesn't have a secret compartment. I decided not to include this as it would have reduced storage capacity and I didn't think that the box would have benefited from it. Other modifications I made were that I used plywood for the base because it increases storage space, it would be easier to affix and because it would be covered with felt you wouldn't notice it is different wood. I also changed the colour of the stain from teak, green and red to just teak. I did this because I had spent four lessons burning the design on the wood and I was afraid the other colours would spoil it.

I also encountered some problems. When using the mitre saw, the clamps put strain on the wood so the saw cut out of line. This meant the pieces had to be sanded down so they sat flush. Because of this, the box is not a regular octagon and some gaps had to be filled with sawdust. Apart from this, there were no other major problems, my jig worked very well and there was no problem with time.

If I were to make this box again, I would probably cut the sides on a band saw instead of using the mitre saw, I would practise different coloured stains and apply them to the dragon and I would include a lock instead of a clasp.

However, I am very happy with the way my box turned out, it works very well and I am pleased with the changes I have made. This box cost around £4 to make and 21 hours to finish. In my view a realistic retail price would be around £10-£15, and it would probably be bought as a present.

BRENDAN ARCHDEACON

26

Evaluation (cont.)

Once my box was finished I tested it to make sure it would be suitable for its function, holding trinkets. As you can see from the picture below, it is big enough and strong enough to hold a wide range of products, and has lots of storage space to hold many other things.

My jig was also successful as my pieces were cut quickly and accurately, with ease. However, because I had to cut the wood again using the mitre saw at an octagonal angle, the jig didn't seem to speed up the making process and I probably would have been better off without it.

To see if my user group liked my product, I asked a few people what they thought of it and here are some of the responses:

'This is a very unique box. I like the way the dragons intertwine but I think it needs more colour.' David Figgins

'I love this elegant box with its oriental lid design. It's also practical and could store all types of trinkets.' Mary Jones

'This box has an oriental charm about it. I like its shape and it would be a great present to give someone.' Sean Payne

From these quotes and my earlier questionnaire results I feel that I have succeeded in making a box for my target user group.

BRENDAN ARCHDEACON

27

EXAMINATION ADVICE

Examination advice

All pupils have to complete coursework (component 5) which is 60% of your final mark. It is therefore important to make the most of this opportunity because you can achieve a good GCSE grade before you take the examination papers. The remaining 40% of the final mark is divided into two examination papers, which carry 20% for each.

Entry options

You and your teacher will decide on what tier you will be entered for. You will be entered for either the foundation or higher tier. The foundation tier examination papers are shorter in length and less detailed than the higher papers but you can only achieve a C grade. The higher papers are longer in length and contain more detailed questions which allow you to achieve an A* to D grade. But if you do not achieve a D at least you will not achieve a grade at all. Therefore it is important to select the correct tier of entry for the examination papers.

If you are entered for the foundation tier you will take examination papers 1 and 3 and the marks will be added to your coursework mark. This will result in your final GCSE grade.

If you are entered for the higher tier you will take examination papers 2 and 4 and the marks will be added to your coursework mark. This will result in your final GCSE grade.

Question papers

Papers 1, 2, 3 and 4 will test your knowledge and understanding of resistant materials through questions on designing and making. There will be no option for selecting questions. You will be expected to answer all of the questions in the papers.

Papers 1 and 2 will include a question on product analysis. This question will relate to the information given in the paper. Papers 3 and 4 on the other hand will include a question on product analysis that relates to a different theme each year. You will be told about the theme earlier in the year so you can prepare for the examination.

Examination tips

Revision

- Write revision notes that are clear and easy for you to understand. Use sketches to help explain your notes.

- Break the notes down into short segments so that they are easy to remember.

- Produce a revision timetable.

- Spend half an hour revising at a time. Have a break for 10 minutes then do another half an hour revision. This will help you to retain information. Don't try to cram it in all at once at the last minute.

Component	Name	Duration	Weighting	Grade
1	Paper 1 (Foundation)	1 hour	20%	C to G
2	Paper 2 (Higher)	1 hour 15 mins	20%	A* to D
3	Paper 3 (Foundation)	1 hour	20%	C to G
4	Paper 4 (Higher)	1 hour 15 mins	20%	A* to D
5	Internal assessment (Coursework)	40 hours	60%	A* to G

- Use mind maps with illustrations to help you visualize processes.

- Go through practice and past examination papers to help familiarize yourself with the layout and wording of questions. This will also test your understanding and knowledge of resistant materials.

- The examination papers contain questions that require you to sketch out ideas and improvements to designs. How you communicate your ideas is very important in achieving good marks. It must be clear to understand. Practice sketching and communication techniques such as 3D and 2D views, detail views, hidden details, annotation etc.

Equipment

You must have a pencil, pen and eraser with you for the examination. It may be useful to bring with you a compass, protractor, set square and ruler to assist you in sketching out ideas. Coloured pencils may also assist in explaining ideas but they should not be used for adding colour to the sketch to make it look better. This will only waste time and you will not receive any marks for a pretty picture.

The paper

- Make sure that you have the correct tiered paper.

- Write your name, centre and candidate number on the paper.

- Read through the instructions and information on the front of the paper. It will have important information that you will need to know before starting the examination such as what the dimensions are measured in etc.

- Quickly check through the paper before you begin. This is to see how many questions there are and if any of the questions have more marks or more time is needed on certain questions.

- Read through the questions twice. Underline the key words so you know exactly what the question is asking for.

- Each question will have the number of marks to be awarded for the answer. Your answer should relate to the marks. For example a 1 mark question would require you to write a sentence to state a point. Depending upon the type of question, a question with 2 marks would require you to write a sentence or more to state two points or a point with a reason for the answer. For 3 marks you will have to write 3 points or a point and two reasons for the answer etc.

- Don't copy out the question in the answer.

- Write your answers in sentences.

- Use the correct technical vocabulary in your answers. This will allow you to explain yourself easier and demonstrate to the examiner that you have a good understanding of the subject.

- All sketches must be in pencil so that you can easily correct them with an eraser.

- Before sketching think about what it is you want to communicate and how you will lay it out on the page. You have limited space so use it wisely.

- If you have finished before the allotted time, check through your answers carefully.

Examples of questions 1

Papers 1 and 2

The picture below shows a toy fire engine made from solid wood suitable for use by children aged 3–6 years.

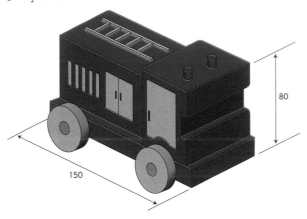

Main features of fire engine:

● painted finish

● cab, chassis, body and wheels permanently attached

● ladders, doors painted on

(All of the following questions relate to the information given here. It is therefore important to examine and understand the information given. The important information is it is made from solid wood, for a child aged 3–6, the design, dimensions and the main features.)

a Name a solid wood commonly used in the manufacture of children's toys.

_____ [1]

(You will need to be specific when stating materials to be used. Just answering with 'Softwood' or 'Hardwood' will not get you a mark. The answer would be beech as it is the most commonly used wood for this type of toy because it is durable, easy to finish and inexpensive compared to most hardwoods.)

b Describe **two** ways in which the design of the fire engine could be considered suitable for a child aged 3–6 years.

1 _____

_____ [1]

2 _____

_____ [1]

(Remember not to include the question in the answer, e.g. 'One of the ways the fire engine could be considered suitable is…'. This is just wasting your time and space for the answer. In examining the information available, what design features do you consider suitable for a child of this age:

● *being the appropriate size?*

● *having a brightly painted finish?*

● *having painted features?*

● *moving wheels?*

Write your answers as a sentence, e.g. 'It is the appropriate size for a child of this age to play with'.)

c State **two** ways in which the designer has considered mass-production in the design of the fire engine.

1_____
_____ [1]

2_____
_____ [1]

(The key words are designer/considered/mass production in the design. How have the shape and features been designed for the ease of mass production? Simple to manufacture shapes, details of windows, ladder etc. applied, spray painted finish.)

d Children's toys can also be made from plastics.
State **two** reasons why consumers would choose to buy a toy made from plastics rather than solid wood.

1_____ [1]

2_____ [1]

(The key words are why/consumer/choose/ plastic/rather than wood. What advantages does plastic have over wood when producing this toy? Inherent colour so it does not have to be painted nor will the paint come off, smooth and rounded parts for safety, intricate detail possible in production, extremely durable etc.)

e Use notes and sketches to show **one** improvement you could make to the design of the fire engine to make it a more exciting toy.

[3]

(The key words are notes and sketches, so you will have to communicate the ideas well; one improvement to the design is required, so do not waste you time on two or three modifications; it must be more exciting, so that the child would want to play with it. Ideas could include:

- *detachable pieces for cab and body etc.*

- *separate ladder which is hinged and lifts, etc.*

To gain 3 marks the modification must be realistic and explained well. This can be done by using different communication techniques, which detail how it will work.

Do not try to spend too much time on this question by sketching an overly complicated idea. It is only worth 3 marks. There will be similar questions in the paper which will have a lot more marks and therefore deserve spending more time on.)

[Total 10]

Examples of questions 2

Papers 3 and 4

Papers 3 and 4 will include a product analysis question based on a theme. Each year the examiners select a different theme to base the question on. You will be told what the theme is earlier in the year so you can prepare for the examination. Below is an example of this type of question.

This question is based on the theme 'Garden Furniture'.

The picture below shows two chairs suitable for use in a garden.

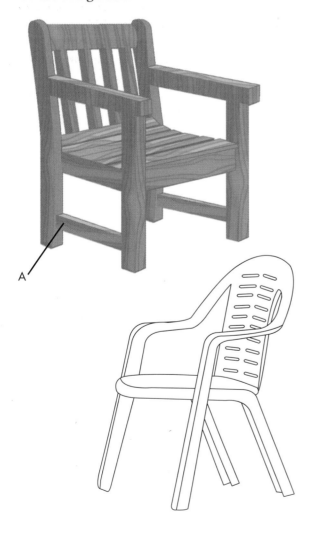

A

a i Name a solid wood suitable for outdoor use.

_____ [1]

(You will have to be specific in the wood you choose, 'Softwood' or 'Hardwood' will not get the mark. Only a wood with specific outdoor properties would be acceptable such as teak, oak, elm, iroko etc.)

ii Name a suitable construction that could be used to join the leg and rail at A.

_____ [1]

(A suitable construction would be the best joint to use for the design. This would be a dowel joint or mortice and tenon joint.)

b i State **two** reasons why a consumer might choose to buy the wooden chair.

1 _____ [1]

2 _____ [1]

ii State **two** reasons why a consumer might choose to buy the plastic chair.

1 _____ [1]

2 _____ [1]

(Putting answers like 'because it looks good' or 'the consumer likes it' will not get you any marks. The question asks for reasons. What advantage does a wooden/plastic chair offer to the consumer? Answers for the wooden chair could be that it:

- *is a traditional material*
- *is durable for outside use*
- *blends in with garden environment.*

For the plastic chair could be that it:

- *is maintenance free*
- *is lighter to move around the garden or to store*
- *is easy to wipe clean*
- *resists weather well.)*

c Explain how anthropometric data would have been used in the design of both chairs.

_____

_____ [1]

(The key words are explain how/anthropometric data/design of chairs. This is a 2 mark question, which asks you to explain your answer so you will have to make more than one point. For 1 mark you will have to say 'To determine the main sizes for the chair' and for the other mark 'By applying specific human body dimensions'.)

d Use notes and sketches to describe **one** improvement you would make to the design of **either** the plastic **or** the wooden chair.

[2]

(In the question it has three words in bold to ensure that you do not do more than it has asked. Only describe one improvement to the design of one of the chairs. If you do more than this you will be wasting time and it is not what the question is asking for.

An improvement to the design does not mean re-designing the whole chair. It will have to be a sensible improvement that is relevant to the design. It might just be restyling the arms and the back or the profile of the legs.)

[Total 10]

Case study 1: Bespoke furniture

Bespoke furniture is produced as a one-off design specifically for a client's requirements. In this case study a client wants a shelving unit that is permanently fixed across one wall to store a collection of books. The client has tried to buy shelving units for this purpose but was unable to. This was because the wall where he wants the unit has two alcoves and the books he wants to store range in size. The client has been unable to find an existing product that is appropriate and fulfils all of his needs. In having a bespoke shelving unit designed and fitted he is making the most of the space available as well as accommodating his collection of different sized books. This one-off design will cost considerably more than buying a pre-manufactured unit.

Before any construction begins, the client meets with the designer to discuss exactly what he wants with regards to construction, material, size and design. The designer will measure the space where the unit is to be fitted and draw up a specification. From this the designer will produce a series of drawings detailing the design of the shelving unit to show to the client. The client can make modifications before production starts. This ensures that the design meets the client's needs and the client is happy with what is going to be produced.

The manufacture of the shelving unit will take place in the manufacturer's workshop. All the parts will be cut and some basic assembly will take place. The manufacturer will try to construct as much of the unit as possible. The final assembly and fitting, though, will have to take place on-site. This allows for easy transportation, but more importantly it will allow for final adjustments to take place so that the shelving unit will fit exactly into the required space. The designer cannot take for granted that the walls will be straight or the floor will be level. When it is fitted, on site alterations can be made to the unit to rectify any problems.

Case study 2: Product testing

Ikea is a large manufacturer and retailer of household furnishings. In order for Ikea to offer quality assurance, they must carry out testing on their products before they are manufactured for sale. Ikea achieve this by:

- carrying out automated testing procedures on their own products

- ensuring that their suppliers of materials, fixtures and fittings have carried out tests which conform to their standards.

If you have ever been to an Ikea outlet you may have seen a demonstration model of the automated tests that are carried out. The presence of this demonstration model on the shop floor is an attempt to show the customer the kinds of testing that are carried out. This reassures the customer by showing awareness of quality standards in their design and manufacture.

Machinery applies repeated pressures to chairs in order to recreate the actions of the user. For example, force is applied repeatedly to the back of the chair and the seat as these are the areas that are most under stress from the weight of the user. Results from this testing will inform the designers where modifications to the design should be made. Similar testing is carried out on other Ikea products to make sure their designs are fit for their purpose.

Case study 3: Large volume furniture manufacturer

Large volume manufacturers of furniture produce a range of existing designs. The range of furniture produced will have been designed for a specific market. In this case study the manufacturer produces traditional solid pinewood furniture. It is primarily aimed to be sold in retail outlets at a low to mid price range.

The manufacturer has several products (lines) in their range, which means that the manufacturer will use batch production methods to produce the furniture. This allows the manufacturer to be flexible in deciding how many products are produced from each line. They can quickly respond to the demand for each line by changing the amount produced.

Each section of the factory will have a production schedule. This lists the quantity of each item that needs to be produced that day. It usually stays the same but it does fluctuate depending upon what is required. The production manager is responsible for producing the schedule and this is given to the team leader for each section.

Before construction can begin the pine needs to be prepared. This involves rough sawing, planing and thicknessing the wood to the required sizes so it can be cut to the correct length. It is then stored on pallets so it can be transported to each section when required.

Due to the fact that pine comes in narrow boards, they will have to be joined to create tops, sides, drawer fronts and doors. In order to do this, the boards are glued and biscuit jointed. Once dry they are placed in a flat bed sander to achieve a level finish and are then cut to the required length.

Once the wood has been prepared and cut to length it can then be machined. This involves three processes:

1 Edge profiles – a spindle moulder (router) is used to round or profile edges.

2 Dowel joints – all of the furniture is made using dowel joints. These joints are strong and can be easily produced using jigs and machinery set up for the task.

3 CAD/CAM – the company use a CNC router to cut out areas for fixtures and fittings such as locks and hinges. The router is also used to produce mouldings on drawer and door fronts.

After the parts have been prepared and machined they are transported on pallets to the assembly section. The parts are then glued and assembled to produce the carcass (the main body of the furniture) and the drawers. These are then sent, along with the doors, to be prepared and sprayed with lacquer. Once dry, all the parts can be finally assembled. In this final section the drawers, handles, doors and fixtures can be fitted onto the carcass. The final product is then packed for protection and stored for delivery.

Case study 4: Injection moulding

The injection moulding process allows low cost products to be produced. The actual initial cost for setting up the production for injection moulding is very high. This is due to the fact that the production of the mould is very expensive. Once a company has invested money into having a mould made, it will have to produce a large quantity of the item over a period of time to make the initial cost worthwhile. This will result in the production of a low cost product because the cost of the tooling has to be included.

The screw tops on plastic bottles are often produced by the injection moulding process. The top must be capable of withstanding repeated use while continuing to keep the liquid inside from spilling. The top also has to have a grip so that the user can open and close the bottle effectively, and it must also be inexpensive to produce.

In producing the top by injection moulding, a degree of precision and accuracy can be built into the design. The mould will allow for a top to have a thread and a grip so it can be easily opened and will seal the bottle efficiently. It will also allow for thousands of tops to be produced without them deteriorating.

The two-piece mould used will produce a number of tops at a time. Channels will have been made for the plastic to run into the mould. Once the plastic has cooled and been removed from the mould, a plastic bead will be connected to the top. This will be trimmed off and as a result the top will have a small nipple. This is how injection moulding products can be identified.

Glossary

acrylic a brittle plastic commonly known as perspex

adhesive a bonding agent used to join two materials

aesthetics how we respond to the visual appearance of a product, in relation to its form, texture, smell and colour

alloys metals made by combining two or more metallic elements; usually to make stronger or more resistant

analyse to study closely and ask questions such as who, what, where, when, why and how

annotation explanatory notes on a design, detailing such things as the materials used, how it works

anthropometrics the study of the human form in relation to size, movement and strength; used in ergonomics

batch production a method of production where a number of components are made all at once. Repeated batches are sometimes made over a longer period of time.

black mild steel steel which has been rolled hot and therefore the surface is left black and slightly rough

blow moulding a process where a thin tube of plastic (parison), gripped between two halves of a mould, is blown out to fill the mould using compressed air. Used for making bottles

brazing spelter an alloy of copper and zinc which melts at 875°C and is used as a filter

brief a short statement of a problem or need

BSI British Standards Institution

CAD Computer Aided Design

CAM Computer Aided Manufacture

casting an object made by pouring liquid metal into a mould and allowing it to solidify

CD-ROM Compact Disc – Read Only Memory

chuck a mechanical screw device for holding a tool in a machine, e.g., a drill bit

clearance hole a hole made to enable free passage of a screw-head

closed loop a system with feedback

CNC computer numerically controlled machines are controlled using number values written into a program; each number is assigned a particular process

continuous production when products are made one after the other

COSHH Control of Substances Hazardous to Health

countersunk hole a hole made to receive the head of a screw and leave it level with the surface

crating boxes used to help 3D sketching

criteria requirements for a product

cutting edge a sharpened edge used to cut

database a collection of information stored on a computer program

deforming process that allows material to change shape without changing its state, i.e. vacuum forming

design the process of solving problems through the development of ideas to produce a solution

designer a person who produces design ideas for clients' needs

development the process of taking an idea and improving/modifying it in order to achieve the best possible solution

die casting a process whereby metals are poured under gravity or injected under pressure into a metal mould

digital camera a device which takes pictures and stores them directly on a computer disk

elevation another name for 'view' on an orthographic drawing

endorsed identity a type of corporate identity

ergonomics the study of how products and environments are designed for human users

evaluation judgements made about ideas and products against the original specification

extrusion the process of forming uniform cross-section of moulding used extensively for plastics and metals

fabrication the process of joining parts together

ferrous containing iron

flotation where substances are separated by floating them in various liquids

flowchart a chart using symbols to show the sequence of a process

flow production continuous production

flux a substance mixed with a solid to lower the melting point; used in smelting and soldering

foamboard a modelling material

former a base used to build up thin layers on a material to produce a desired shape or curve

function what a product or process is expected to do

Gantt chart a chart to show how a number of tasks/processes can be completed in a given time, often simultaneously

gauge a standard measure of the thickness of a screw, wire, sheet metal, etc.

hardness the ability to withstand abrasive wear and indentation

hardwood wood from a broadleaved tree

horizon line the line of sight on a perspective drawing

injection moulding a process where molten thermoplastic is injected under high pressure into a die cavity

isometric a method of drawing objects in three dimensions using 30-degree axes

jig movable holding device made to suit a single component in exact position

justify to give reasons for choices

just-in-time an industrial method of stock control

laminating the process of joining sheet materials together to form solid sections or curved shapes

manufacturing cells a production system that incorporates a number of people and machines working together, being responsible for what is produced

marketing the selling of a product or service to a consumer

mass production the production of a component or product in large numbers

MDF Medium Density Fibreboard

mechanical properties properties of materials that are effected by an external force such as compressive or tensile forces

mechanical strength materials which are said to be strong, and resist or stand up to external forces such as compressive and tensile forces

MIG welding Metal Inert Gas welding – a relatively easy form of welding to carry out, commonly used in many school workshops

mock-up a model of a design in 3D, used for evaluation and testing

non-ferrous a metal that does not contain iron or steel

one-off production a product required as a single item, such as a bridge or a football stadium

open loop a system with no feedback

ore a natural material from which metals or minerals can be extracted

orthographic a form of technical drawing where an object is drawn from different views

perspective the appearance objects give of being smaller the further away they are

physical properties properties other than those which are effected by an external force such as density and conductivity

pilot hole a small hole used to guide a screw or thread

planed all round (PAR) timber that has been planed all round will have four smooth machined surfaces

plotter/plotter cutter a computer controlled output device for producing accurate lines or cuts on card or paper

presentation drawings drawings used to communicate a design in suitable form for the client

prototype a model or product which has been made to be tested or trialled before being put into full production

qualitative information based on opinion or observation

quality assurance a policy or procedure written to ensure that a product reaches the customer to the correct specification

quality control systems put into place to check quality during manufacture – e.g. gauges, visual checks, etc.

quantitative data that can be measured

questionnaire a survey made up of related questions which help to find out people's views

realization taking a design on to the next stage after planning, that is, making a real product

reforming process involving a change of state within the materials used, i.e., from solid into liquid

rendering applying colour or texture to a drawing

research the gathering and analysing of information

rough-sawn timber which has come straight from the saw and has not been planed

scale representing dimensions on a drawing in proportion which can be greater or smaller

scale of production the type of production – batch, mass, etc.

scanner a device that produces a digital image

seasoning the process of reducing the moisture content of timber

section an orthographic view showing a cut-through view

shape memory alloy (SMA) an alloy that can be plastically deformed at a predetermined temperature, but that will return to its original shape after it has been heated

sketch a freehand drawing

smelted metal extracted from its ore by heating and melting processes

softwood wood from a cone-bearing (conifer) tree

soldering joining less fusible metals with solder (a low melting alloy)

specification the criteria that the final solution must achieve

structural members the individual components/parts that are subjected to forces, for example chair legs being put in compression

survey research carried out by questioning

swarf fine filings of metal produced, for example, when cutting screw threads

synthetic product made by chemical synthesis; usually to imitate a natural product

system a group of processes organised to perform a task

target market the type of people or age group a product is aimed at

template a device used to mark out identical shapes

tensile able to be drawn out or stretched

testing checking the outcome against the specification

thermoplastic a type of plastic which softens under heat and can be resoftened many times

thermosetting plastic a type of plastic which, once set, cannot be resoftened or melted

tolerance the upper and lower limits of a dimension, e.g. 25mm+/-0.5mm=24.5/25.5mm

torsional forces forces which work to untwist materials or structural members and components

turning the process of removing or wasting wood and metal

user evaluation a method of evaluating a product by asking intended users what they think of the design and recording the results

vacuum forming a process using a thin plastic sheet which is formed around a mould using atmospheric pressure – used in blister packaging

vanishing point in perspective drawing where construction lines meet

veneers thin strips of hardwood (0.5mm thick) used for laminating cheap material to improve visual appearance

wasting process that produces unusable material by either cutting pieces out or cutting pieces off, e.g., chiselling

web page/site information page on the Internet

World Wide Web another name for the Internet

Index